15 Dangerously Mad Projects for the Evil Genius™

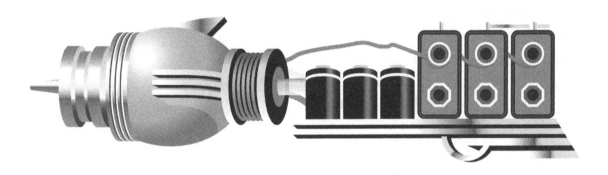

Evil Genius™ Series

15 Dangerously Mad Projects for the Evil Genius™

Simon Monk

New York   Chicago   San Francisco   Lisbon   London   Madrid
Mexico City   Milan   New Delhi   San Juan   Seoul
Singapore   Sydney   Toronto

The McGraw·Hill Companies

Library of Congress Cataloging-in-Publication Data
Monk, Simon.
  15 dangerously mad projects for the evil genius / Simon Monk. — 1st ed.
    p. cm. — (Evil genius)
  Includes index.
  ISBN 978-0-07-175567-2 (pbk.)
  1. Electronic apparatus and appliances—Design and construction—Amateurs' manuals. I. Title.
II. Title: Fifteen dangerously mad projects for the evil genius.
  TK9965.M66 2011
  621.381—dc23

2011014332

McGraw-Hill books are available at special quantity discounts to use as premiums and sales promotions, or for use in corporate training programs. To contact a representative, please e-mail us at bulksales@mcgraw-hill.com.

15 Dangerously Mad Projects for the Evil Genius™

1 2 3 4 5 6 7 8 9 0 QDB QDB 1 0 9 8 7 6 5 4 3 2 1

ISBN 978-0-07-175567-2
MHID   0-07-175567-5

Sponsoring Editor
  Roger Stewart

Editorial Supervisor
  Jody McKenzie

Acquisitions Coordinator
  Joya Anthony

Project Manager
  Patricia Wallenburg

Copy Editor
  Mike McGee

Proofreader
  Claire Splan

Indexer
  Claire Splan

Production Supervisor
  George Anderson

Composition
  TypeWriting

Art Director, Cover
  Jeff Weeks

Cover Illustration
  Todd Radom

To my mother, Anne Kemp, whose love, kindness, and good cheer
continue to serve as an example to me and all who know her.

About the Author

Simon Monk has a bachelor's degree in Cybernetics and Computer Science and a doctorate in Software Engineering. He has been an active electronics hobbyist since his school days and is an occasional author in hobby electronics magazines. He is also author of *30 Arduino Projects for the Evil Genius*.

Contents

Acknowledgments

I THANK LINDA FOR GIVING ME the time, space, and support to write this book and for putting up with the various messes my projects create around the house. You are a gem!

I also thank my boys, Stephen and Matthew Monk, for taking an interest in what their Dad is up to and their general assistance with project work.

Finally, I would like to thank Roger Stewart, Patricia Wallenburg, Mike McGee, Joya Anthony, and everyone at McGraw-Hill, who did a great job once again. It's a pleasure to work with such cultured and enthusiastic individuals.

Introduction

THIS IS A BOOK OF PROJECTS. They are projects designed to appeal to the Evil Genius in everyone, whether those persons are new to home construction or experienced project-makers.

Most projects require some knowledge of electronics and the ability to solder. Two projects—the trebuchet and the ping-pong ball minigun—have nothing to do with electronics whatsoever. A handful of the projects, like the persistence-of-vision display, require the use of an Arduino microcontroller module. Some of the projects can be completed in an evening, while others may take several weekends to finish.

What all the projects have in common is that they are all, in some way or another, at least a little "dangerously mad" and have a certain "wow" factor. So whatever your skill level, there should be a project for you in this book.

Each project includes a parts list and step-by-step instructions, accompanied by numerous diagrams and photographs. There is a "Theory" section at the end of most chapters that gives background information about how the projects work. So, even if you are new to construction, you should find that if you start with some of the simpler projects, you will soon feel confident enough to tackle some of the more difficult ones.

Ratings

To give you a clue as to what you are in for before starting a project, each project is rated according to the amount of time it will likely take and the level of technical expertise required.

So it should be possible to complete a project rated as "Small" in half a day to a day. A "Medium" project will probably take you a weekend, and a "Large" project maybe more than one weekend. Your mileage may vary—and as is always the case with such things, it will depend very much on whether things go smoothly or not.

The star ratings for difficulty are as follows:

★ Does not require any soldering or any actions more complex than measuring, cutting, drilling, and gluing.

★★ A small amount of soldering is required, but no circuit boards will need to be made, and no microcontrollers must be programmed.

★★★ A simple design will need to be made, which may use a small stripboard. You should be reasonably proficient at soldering and able to use a multimeter.

★★★★ A complex design must be constructed, requiring the ability to solder a stripboard. Also, some mechanical construction is necessary.

The Web Site

A web site accompanies this book (www.dangerouslymad.com), where you will find:

- Source code for the projects that use the Arduino microcontroller

- A message generator for the persistence-of-vision display in Chapter 8

- Videos and photographs of the completed projects

- Ways of contacting the author

- Errata

Arduino

Three of the projects in this book use Arduino microcontroller boards. These readymade boards offer a very easy way of using a microcontroller. They have a USB connection for programming and are accompanied by an easy-to-use development environment.

Chapter 8 contains the first project to use an Arduino board and includes step-by-step instructions for the installation and setup of the Arduino software.

For more information about Arduino, please refer to the official Arduino web site, www.arduino.cc, or you may wish to buy *30 Arduino Projects for the Evil Genius*, also by this author.

Electronic Construction

The Appendix contains a primer on the basics of electronic construction. If you are new to electronics, you should find this a useful resource, and you should probably read it before you embark on one of the projects involving electronics.

Getting Started

So, it's time to get started. Browse through the following projects and see which one inspires your inner Evil Genius.

A Word About Safety

The projects in this book are meant to be educational and fun, but they also require that you exercise a reasonable degree of caution. When working with tools and electricity, you should always make sure that you have the proper training, take the proper precautions, and wear the appropriate safety equipment. The projects are designed for qualified adults and should not be attempted by children without adult supervision. Lasers and projectiles should never be aimed at people or pets. McGraw-Hill and the author do not warrant the safety or fitness of any of the final products resulting from the projects. Your safety is your responsibility.

EVIL GENIUS PROJECT DESCRIPTIONS AND RATINGS				
Chapter Number	**Project**	**Description**	**Size**	**Skill Level**
1	Coil Gun	Gun that fires small lengths of nail	Medium	★★★
2	Trebuchet	A simple version of the medieval siege engine	Small	★★
3	Ping-Pong Ball Minigun	Gun that fires ten balls per second at 30 mph or more	Small	★
4	Mini Laser Turret	A model remote control laser	Medium	★★★★
5	Balloon-Popping Laser Gun	A clever bit of trickery that pops balloons and makes cans jump when hit by a laser	Large	★★★★
6	Touch-Activated Laser Sight	A laser that comes on when you touch the trigger	Small	★★★
7	Laser-Grid Intruder Alarm	A high-tech alarm that bounces laser beams between mirrors	Medium	★★★
8	Persistence-of-Vision Display	Motorized multicolored LEDs that write in the air	Large	★★★★
9	Covert Radio Bug	A short-range bug that uses an adapted MP3 radio transmitter	Medium	★★★
10	Laser Voice Transmitter	A device that sends your voice over a laser beam	Large	★★★
11	Flash Bomb	A modified single-use camera that flashes if it's picked up	Small	★★
12	High-Brightness LED Strobe	A powerful strobe light (even an Evil Genius likes to party)	Medium	★★★★
13	Levitation Machine	A microcontroller-based electromagnetic suspension machine	Medium	★★★
14	Light-Seeking Microbot	A tiny light-seeking robot	Small	★★★★
15	Surveillance Robot	An autonomous robot that roams about looking for intruders and sounds an alarm when it finds them	Large	★★★★

CHAPTER 1

Coil Gun

PROJECT SIZE:	Medium
SKILL LEVEL:	★★★☆

THIS COIL GUN (Figure 1-1) will fire a small metal projectile at up to 30 miles per hour. It is portable, being powered by batteries, and is guaranteed to strike fear into the enemies of the Evil Genius. On dark evenings, the Evil Genius likes to strap flashlights to the heads of his minions and make them run around the Evil Genius' Lair while he takes potshots at them. Oh, how they squeal in panic!

A coil gun works in a similar way to a photographic flash gun. Capacitors are charged up over a few seconds, and then all their electrical energy is released extremely quickly. In a flash gun, the energy is released through a flash tube, and in a coil gun it is released into a coil of wire. This creates a powerful magnetic field that will cause any iron object near the coil to move.

Since the coil is wrapped around the tube from a plastic pen, and the iron projectile is inside the tube, it will fly along the tube towards the coil. As all the energy from the capacitors will be spent in a matter of milliseconds, the coil should ideally be turned off by the time the projectile passes its center and exits out the other side of the tube.

A similar, but less attractive analogy to the coil gun is a toilet cistern. In this case, the tank is like the capacitor, except that it is filled with water rather than charge. The tank charges over a period of a few tens of seconds. When the toilet is flushed, all the water rushes out.

The gun is controlled from a single three-position switch. When in its center position, the gun is off. When pushed forward, it starts to charge and the charging LED comes on. When fully charged, the LED goes off and the gun is ready for firing by pulling the switch back like a trigger.

Figure 1-1 The coil gun

WARNING!

This is not a real gun. You could probably throw the projectile as fast as it comes out of the end of the gun. In addition, the projectile is lightweight.

However, this project has a number of dangerous aspects.

High currents. These are not high voltages, but the currents are very high and produce strong magnetic fields, so do not build this project or use it anywhere near anyone with a pace-maker.

Do not short-circuit the capacitors when they are charged up. You may melt whatever you are shorting them with, which means there will be small quantities of molten metal flying around.

And do not place your eye or anyone else's in the line of fire of this gun.

What You Will Need

The components for this coil gun are all readily available. It's worth shopping around for the capacitors; eBay usually has a good selection of suitable ones if you search for "electrolytic capacitors." You'll need the parts in the Parts Bin.

You will also need the following tools listed in the Toolbox.

TOOLBOX
■ Soldering equipment
■ Hacksaw
■ Wood saw
■ Drill and assorted drill bits
■ Epoxy resin glue or hot glue gun
■ Multimeter

PARTS BIN			
Part	**Quantity**	**Description**	**Source**
Firing tube	1	Disposable transparent ballpoint pen about ⁵⁄₁₆ of an inch (7mm) in diameter	
Coil retainers	2	Plastic brackets cut from a plastic food container (see the following description)	
Projectiles	2	Iron/steel nails ⅛ inch (3mm)	Hardware store
Lumber		18 inches (45cm) length of ⅝" × 1¼" (18mm × 33mm) wood	Hardware store
Plastic drink bottles	1		
Plastic insulating tape			Hardware store
Coil wire		13 feet (4m) of 20 or 21 AWG enameled copper wire	Farnell: 1230984
Batteries	4	Budget PP3 9V batteries	
Battery clips	4	PP3 battery clips	Farnell: 1183124
C1-8	8	4700µF 35V electrolytic capacitors, or any set of 35V capacitors totaling around 38,000µF	Farnell: 9452842 eBay
SCR	1	40TPS12A Thyristor 55A	Farnell: 9104755
S1	1	DPDT On-off-momentary toggle switch	Farnell: 9473580
R1	1	100Ω 2W	Farnell: 1129029
R2	1	100Ω 0.5W	Farnell: 1127903
D1	1	5mm Red LED	Farnell: 1712786
R3	1	2.7KΩ 0.5W	Farnell: 9338667
D2	1	5.1V 5W Zener diode	Farnell: 1705663

Assembly

The schematic diagram for the project is shown in Figure 1-2. Only a few components can be soldered together without the need of a circuit board.

The design has three main sections: the capacitor bank, the trigger circuit, and the charge

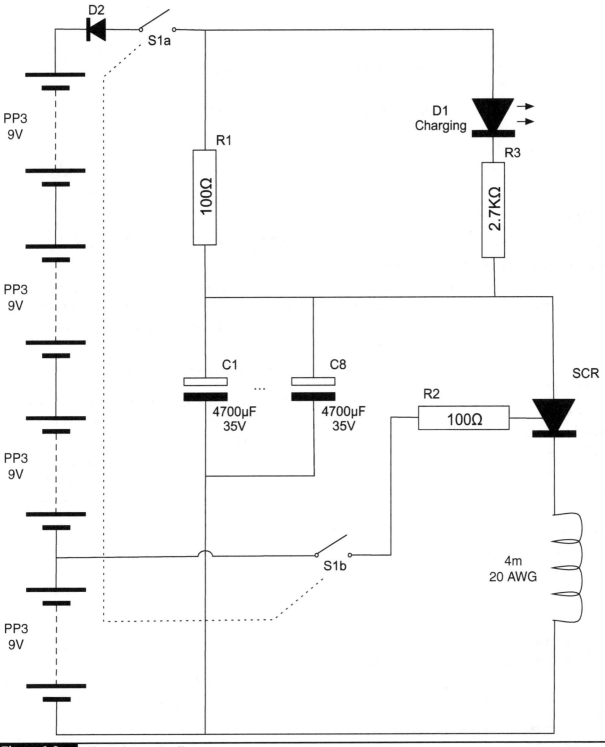

Figure 1-2 The schematic diagram

indicator LED. The capacitors are all connected in parallel and charged by the batteries through the 100Ω resistor when the switch is in the "charge" position.

The trigger circuit uses a SCR (silicon-controlled rectifier), or Thyristor as they are sometimes called. The SCR acts as a conducting switch when a current passes through its "gate" connection. This happens when the toggle switch is put to its "fire" position.

Figure 1-3 shows the full wiring diagram for the coil gun. To maintain a "gun-like" shape, the components are laid out in a line, with the batteries at the back and the coil at the front.

Step 1. Make the Coil Former

The coil is wound onto a disposable ballpoint pen tube. The narrow end of the tube, where the nib would go, should be sawn off and the bung removed from the other end of the tube. To hold the windings of the coil in place, we use right-angle plastic brackets cut from a food container (Figure 1-4). Any kind of plastic with a 90-degree bend can be used here.

The brackets should be about 1 inch (25mm) on each side. They are then drilled in the center so that they fit snugly over the tube. Drill a small hole on one of the brackets, immediately adjacent to the large hole (Figure 1-5). This is where the inner-most connection to the coil will emerge, so the hole must be just big enough for the coil wire. The brackets should then be fixed in place, leaving a

Figure 1-4   Bracket to hold the firing tube

gap of about ⅜ of an inch (10mm) between them (see Figure 1-6). Epoxy resin glue or glue from a hot glue gun is used to hold the discs in position.

Step 2. Wind the Coil

The coil is made up of 13 feet (4m) of 20 AWG enameled copper transformer wire. Winding coils by hand is a little tedious. Fortunately, this coil uses a short length of wire. It is worth trying to wind the coil neatly, but it usually goes off the rails as you get toward the end. This does not really matter.

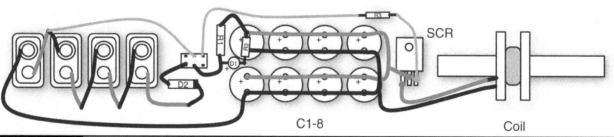

Figure 1-3   The wiring diagram

Figure 1-5   Drilling the bracket

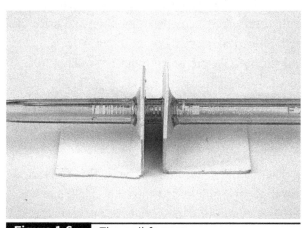

Figure 1-6   The coil former

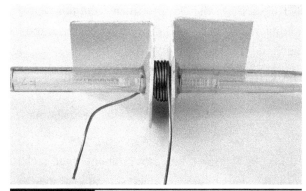

Figure 1-7   Winding the coil

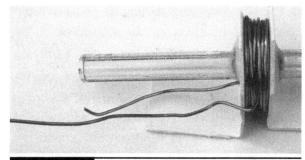

Figure 1-8   The finished coil assembly

Start by measuring out 13 feet (4m) of wire. Thread about 2 inches (50mm) of the wire through the small hole made in the disc, close to the pen tube. This will be one lead of the coil. Then coil the wire around the pen, keeping each turn as close as possible to the previous turn (Figure 1-7).

When you get to the disc at the other end, keep winding in the same direction but allow the turns to line up back toward where you first started. Try to keep the turns as close together and as tight as possible. It can help to, from time to time, put a drop of superglue onto the coil to hold the turns into place.

You should end up with about seven layers on the coil (Figure 1-8). Leave about 2 inches (50mm) free and carefully cut a slot in the edge of the bracket for the free end of the coil. Then, add a bit more glue to make sure the coil stays together.

Later on, we will need to solder the ends of the coil, so scrape the enamel off the ends of the wire and coat the ends with solder.

Step 3. Assemble the Capacitor Bank

The capacitors used in this project were selected to provide the most farads per buck. We used 8 × 4700μF (microfarad) capacitors rated at 35V. This gave us a total of 37,600μF. Four 10,000μF capacitors will work just as well, if not slightly better, when it comes to holding a little more charge. However, you must make sure that the voltage rating is 35V or more.

You should also avoid the temptation to greatly increase the capacitance, as this will increase the maximum current, which may be too much for the

SCR. You may wish to experiment with this, but do understand that when the current becomes too much, it will destroy the SCR.

Figure 1-9 shows how the capacitors are connected together into two rows. It is easiest to make each row of four first and then connect the two rows together.

Start by lining up four capacitors on their backs with their legs in the air. Make sure that all the negative leads are on one side and all the positive on the other. It's very important that the capacitors are connected the right way around. If one of the capacitors is the wrong way around, it could explode—and capacitors are full of messy goo!

Now take some solid core wire and connect all the negative leads together, and then do the same for the positive leads. You can use the same wire as you used to wind the coil, but you will need to scrape away the insulating enamel where you want to make a solder joint. I used solid core wire of the type employed in domestic electrical wiring. This has the added advantage of being able to use the plastic insulation to color-code the positive and negative connections to the capacitor bank.

Use the thickest wire you can get your hands on. This wire is going to carry a current of around 100A and the thicker the wire, the lower the resistance and the more energy will be transferred into the coil.

When both rows of four capacitors are complete, you need to join the common positive connection of one bank to the common positive of the other bank. Do the same for the negative connections (refer back to the wiring diagram of Figure 1-3).

Step 4. Add the Triggering SCR

You may be wondering why we need to use a SCR and why we couldn't just use the switch directly between the capacitors and the coil. The answer is that no regular switch would withstand the hundreds of amps that flow when the coil is triggered. It would simply weld the contacts together or melt them.

The SCR that we have chosen is a good compromise between power handling and price. It will allow peak currents of up to 500A for a millisecond. We are going to need it to handle about 100A for 10 milliseconds.

The SCR sits between the capacitors and the coil (Figure 1-10). The 100Ω resistor is connected to the gate connection.

The middle connection to the SCR is connected to the positive side of the capacitor bank, and the leftmost connection to one side of the coil. The other side of the coil is connected to the negative side of the capacitor bank (see Figure 1-10).

Figure 1-9   Constructing the capacitor bank

Figure 1-10   The SCR and gate resistor

Step 5. Fitting the Batteries and Switch

The four 9V batteries are connected in series to give a total of 36V. However, fresh batteries may have upwards of 10V per battery (a total of 40V), which is above the rated voltage of the capacitors. To play it safe, the Zener diode in series reduces this by about 5.6V, bringing the voltage just below the capacitors' rated voltage. Note that exceeding the rated voltage of electrolytic capacitors is dangerous and will shorten the life of the capacitors.

Cut the battery leads so they are a more manageable length and then connect the positive (red) lead of the first battery lead to the negative lead (black) of the second lead, and so on. Finally, connect the Zener diode in series between the last positive lead and one side of the switch, as shown in Figure 1-11.

The switch is what is called a double-throw switch (Figure 1-12). That is, it is actually two switches operated by a single lever. One of the "throws" of the switch is used to turn charging on and off, and the other is used as a trigger. We now need to connect the "trigger" throw of the switch

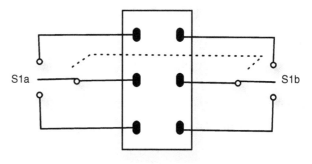

Figure 1-12   Double-throw switch

to R3, which we have already soldered to the gate of the SCR (see Figure 1-3). The other connection to the trigger throw of the switch should be connected to the positive connection of the first battery. For convenience, that can be where the battery leads of the first and second batteries are soldered together.

The switch used has three positions. The center position turns both halves of the switch off. This is the position shown in Figure 1-12. Push the toggle one way and the switch will latch on. This will be the charging position. Pulling the toggle lever the other way will connect the switch momentarily, but the switch is sprung to pull the switch back to the center off position.

So, when connecting the switch you need to make sure it is the right way around so that the firing circuit is switched when it goes into its non-latching action and the charging circuit is made when the switch is in its latched mode. When the switch toggle is in the up position, this usually connects the center connection to the bottom pin, and vice versa. If you get it wrong the first time, you will find you have to hold the toggle in the momentary position to start charging. If this is the case, just unsolder all the leads to the switch and flip it through 180 degrees, then solder the leads up again in the same positions as they were before.

Figure 1-11   The batteries and switch

Step 6. The Charging LED

The charging LED will light when the capacitors are still filling with charge. Once they are full, it will go off and the gun will be ready for firing. It is just an LED with a series resistor to limit the current. It will start bright and gradually get dimmer as the capacitors get fuller.

Figure 1-13 shows the LED and resistor soldered across the charging resistor. Note that the positive (slightly longer) lead of the LED must be connected to the battery end of the charging resistor, and the negative end must be connected to the LED's current limiting resistor.

We have now soldered everything together, but before we start fitting things into a case, we need to test out our coil gun on the bench.

Step 7. Making the Projectiles

We need something for our coil gun to fire. Iron nails ⅛ of an inch (3mm) in diameter are good for this, but they are a little long. Our projectiles should be about ¼-inch (5mm) long (Figure 1-14). Since, these can be hard to find once fired, it is a good idea to have a few.

The Evil Genius has discovered that the best way to find lost projectiles is to take away the shoes and socks of his minions. While walking about barefoot, the fragments of nail invariably attach themselves to their feet.

Figure 1-13   The charging LED and resistor

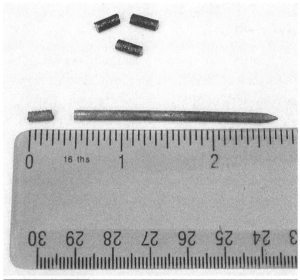

Figure 1-14   The projectiles

Use a hacksaw to cut the nail into pieces the right length.

Test Firing

Now we get to the exciting bit! Before we start, we need to check that everything is as it should be. We are using very high currents here, so there is the potential to destroy our components if we are not careful.

Basic Checks

Before connecting the batteries, compare all the wiring with Figure 1-3 and make sure we have all our connections right. Once you are sure everything is OK, put the switch into its center off position and connect up the batteries.

We are going to start with a low voltage test before we ramp up to full power. Put your multimeter onto its 20-volt range, or at least enough to display 10V, and then connect the multimeter leads across the capacitor bank, as shown in Figure 1-15.

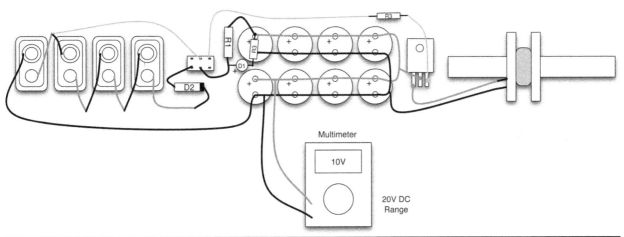

Figure 1-15 Measuring the charge voltage

Put the switch to the "charge" position. If the LED does not light, put it immediately back to the center off position and check your wiring. Also, check that your switch is wired the correct way around. You can test this last point by just putting the switch to the momentary "fire" position. If the LED lights, then your switch is probably the wrong way around (see the earlier section titled "Step 5. Fitting the Batteries and Switch").

Note that, during charging, R1 will get hot.

As soon as the multimeter indicates about 10V, put the switch back to the center position. Watching the multimeter, you should see the voltage decrease very slowly. If it drops down to 0V quickly, then something is wrong, so put the switch back to off and check everything.

Assuming this test passed, we can now push the projectile into the firing tube until the end is just level with the edge of the coil on the far side of the coil. Now put the switch to fire and the projectile should move, hopefully traveling through the tube and emerging at a modest speed.

Congratulations! Since everything seems in order, it's time to try a full-power test. This time, you can just let the gun charge until the LED turns off, which should be after 10 or 15 seconds.

Measuring the Projectile Speed

You can tell if the firing was a good one, and the projectile was going fast, just by observation, but it's better to have a more precise way of measuring the speed. The way to do this is with a humble computer. I am indebted to the excellent Barry Hansen for his web site at www.coilgun.info that, amongst many other useful things, describes the decidedly Genius (I would not say Evil) way of measuring the speed with nothing more than a computer with a microphone input.

Here is how it works. You just record the sound of your test and then use some sound software to examine the waveform and measure the time between the coil firing and the projectile hitting its target. I think it's safe to say that our projectiles are slow enough that we can ignore inaccuracies due to the speed of sound.

First of all, you will need to download some sound recording software. I used Audacity (http://audacity.sourceforge.net) because it is free and available for most operating systems, including Windows, LINUX, and Mac.

You then need to set up your coil gun a convenient distance away from a target that will make a noise when the projectile hits it—say, about 3 feet (1m). I use a plastic shopping bag hung from a door handle that is conveniently the

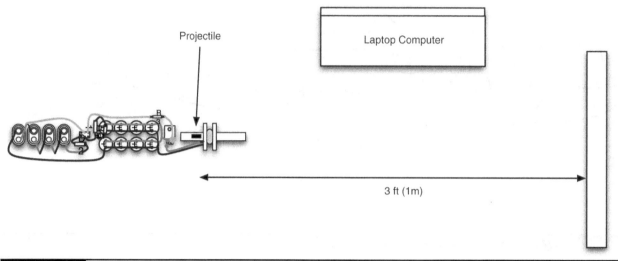

Figure 1-16   The test firing arrangement

same height as my workbench. The bag has the added advantage of absorbing most of the energy from the projectile, thus preventing it from bouncing off and becoming lost. This arrangement is shown in Figure 1-16.

Once your capacitors are fully charged, the LED turns off, and the voltage across the capacitors is about 35V, place the projectile just to the far side of the coil from the target, start your sound software recording and fire the gun. As soon as the gun has fired, stop the recording and look at the sound waveform.

After a bit of cropping and zooming on the resulting sound waveform, you should see something like Figure 1-17.

The human ear is a wonderful thing, and you can use yours to check that the sound spikes correspond to the firing and target impact by selecting one of the areas and clicking play to hear what it sounds like. From Figure 1-17, you can see that the time from the end of the sound of the triggering to the first sound of the impact is 0.15 − 0.085, or 0.065 seconds.

You now need to measure the distance from the starting position of the projectile to the front of the target. We can calculate the velocity as the distance divided by the time. In this case, the distance was

1m and the time 0.065s. So the velocity was 15.39 meters per second. Multiply this by 2.237 to get a figure of 34.4 miles per hour.

I found it best to make a spreadsheet for the results, as shown in Figure 1-18. Successive test results can each be recorded on a separate row.

We now need to find the optimal starting position for the projectile. Using a thin pen that will write on the firing tube, draw a series of dots a few millimeters apart along the tube from the far side of the coil from which you want the projectile to emerge. Then, using the same projectile each time, measure the velocity at each of the positions. You can either line up the front or the back of the projectile with the dot. It does not matter as long as you are consistent.

Putting the Project in a Case

At the moment, our gun is not very portable, so we need to build a case for it. The components look quite impressive, so we are going to make a transparent case, but mount it onto a strip of wood (Figure 1-1). The plan for the wood is shown in Figure 1-19. There are just two bits of wood, and the exact sizes are not critical. The main piece needs to be long enough to accommodate all the components laid out in a line, and the end piece

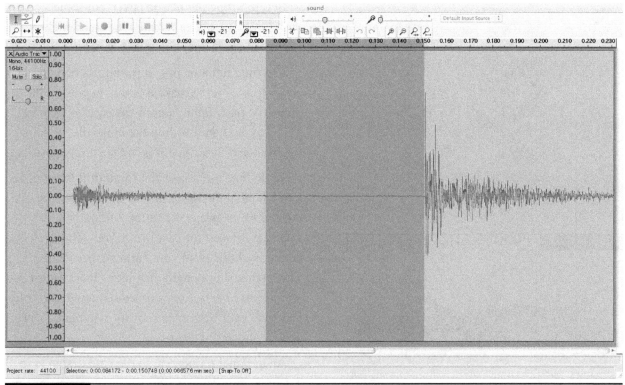

Figure 1-17 The sound waveform from a test firing

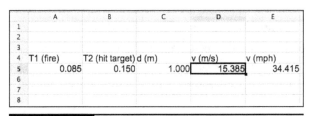

Figure 1-18 A spreadsheet to calculate the velocity

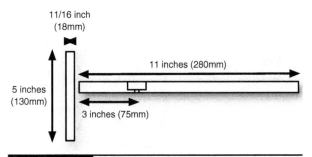

Figure 1-19 The wooden frame for the coil gun

serves the dual purpose of providing a handle and keeping the batteries in place. The end piece is drilled and screwed to the main piece.

Two holes should be drilled into each of the brackets of the firing-coil assembly so they can be screwed into the wood.

Figure 1-20 shows a close-up of the hole for mounting the switch. Drill most of the way through the wood—leave about 3⁄16 of an inch (5mm) with a 1 3⁄16-inch (30mm) bit, which should make a hole large enough to accommodate the whole switch. Then, drill the remainder of the hole through to allow the neck and toggle of the switch to push all the way through. Afterward, fasten the switch on with its retaining nut.

This requires a bit of care to make sure you do not drill the bigger hole all the way through the wood.

Figure 1-21 shows the switch fitted like a trigger below the electronics, as well as the

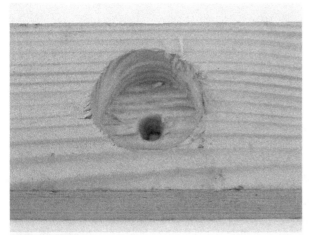

Figure 1-20   The hole for mounting the switch

Figure 1-21   The switch mounted as a trigger

arrangement of the capacitors and other components on top of the wooden structure.

The top view of the whole project is shown in Figure 1-22. Note how we have used a little brass hook to keep the batteries in place. You should also use self-adhesive pads or dabs of glue from a hot glue gun to keep the capacitors in position.

Before fitting the plastic cover, put some insulating tape over any bare wires that might move and touch something they shouldn't.

The final refinement—done to make the gun easier to use—is to make a "stop" to prevent the projectile from falling out of the back of the coil. This way we can just drop the projectile in from the front and know that it is in the correct position.

To do this, use a $\frac{1}{32}$-inch (1mm) drill bit to make a tiny hole in the firing tube at the position where the end of the projectile furthest from the coil should be placed for best firing. You will have determined this from your earlier experiments. Then, put a short length of $\frac{3}{8}$-inch (about 1mm) wire (I used a bit of resistor lead) through the hole and bend over both ends so it stays in place. You can see this in Figure 1-23.

An optional refinement to the case design is to cut some thin Perspex or other flexible plastic (say, from a large plastic bottle) and bend it over the wood from one side to the other, fixing it in place with screws.

Simply take a plastic drink bottle and cut off both ends, then measure out the right length and width of the curved bottle plastic to fit round the top of the gun. Refer back to Figure 1-1 to see how this looks.

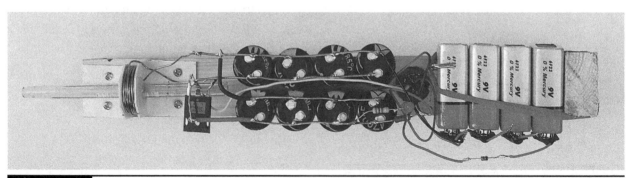

Figure 1-22   Top view showing everything in place

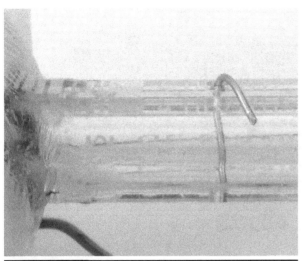

Figure 1-23 Close-up of the end stop

Theory

Coil guns are remarkably inefficient. You are doing well if 1 or 2 percent of the energy in the capacitors is converted into kinetic energy of the projectile. If you can weigh your projectile, or perhaps weigh a few of the nails it is made from, and then do the math, you can calculate the efficiency.

The energy stored in the capacitor is calculated by the formula:

$$E = (CV^2)/2$$

Where E is the energy in joules, C is the capacitance in farads, and V is the voltage.

So, for our arrangement of capacitors, the energy available for each firing is:

$$(0.0376 \times 352) / 2 = 23 \text{ joules}$$

This is similar in energy to the kinetic energy in a .177 air gun. So if we had 100 percent efficiency, our coil gun would be really quite dangerous.

So, now that we know that the "input" energy is 23 joules, let's see how many joules of kinetic energy there are in the projectile.

To do this, we can use the formula:

$$E = (mv^2)/2$$

Where m is the mass of the projectile in kilograms, and v is the velocity in meters per second.

My projectiles weigh approximately 0.3 g and the best velocity I got was 15m/s. This velocity is actually the average velocity of the projectile's flight to the test target, which will be a little less than the muzzle velocity, but it is probably close enough not to matter much.

So, the energy transferred from the capacitors to the projectile is:

$$(0.0003 \times 15 \times 15)/2, \text{ or } 0.033$$

This means that the efficiency of our coil gun is:

$$0.033/23, \text{ or } 0.14 \text{ percent}$$

Not brilliant. So where does all the energy go?

Some of the energy is lost because the projectile is first pulled into the coil, and then if the magnetic field has not disappeared before it passes the center of the coil, it will be pulled back into the coil, slowing it down. The answer to this is to shorten the pulse, which can be done in two ways:

■ Use fewer capacitors (but this will reduce the input energy).

■ Use less turns of wire for the coil (but this will increase the maximum current, potentially destroying the SCR).

The geometry of the coil and the type and size of the projectile, as well as the material it is made from, all affect performance. Getting the most out of your coil gun is very much a matter of trial and error and reading about other people's trial and error.

The Internet offers some great resources on this. The Wikipedia entry for coil guns is a great starting point, as is www.coilgun.info.

If you have an oscilloscope, you can use it to measure the duration of the pulse through the coil. You will need to set your scope as follows:

- **Channel sensitivity:** 5V/div. Make sure the scope is okay with 35V.

- **Trigger mode:** single shot, rising edge.

- **Timebase:** 2.5 ms/div.

You can see a firing of the gun in Figure 1-24. This shows that at 2.5 ms/division, the pulse is about 15 ms in duration.

Summary

This is a fun project to make. As we discovered in the theory section, the gun is very inefficient. This is probably just as well, since it could do some damage if it operated at anything like 100 percent efficiency.

Improving the efficiency at least to one or two percent is quite possible, but will absorb quite a lot of your time.

To do this, you will probably want to buy a spare SCR and test the one you have to destruction with a shorter coil using thicker wire. You may also want to experiment with bigger and better capacitors, or a higher voltage circuit.

You can find much higher power SCRs, but they will be expensive, unless you get them from a surplus store. Certain online stores sell surplus

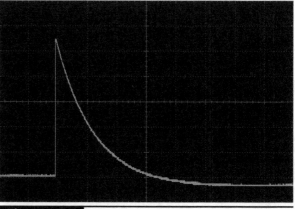

Figure 1-24   Oscilloscope trace of a test firing

electronics, and eBay usually has a selection of high-power SCRs

You may find that taking apart the soft iron laminations from a transformer and layering them around the coil will increase the efficiency considerably.

Whatever techniques you decide to employ, be careful, and expect to destroy the occasional SCR. As well as being cautious, you need to be scientific in your approach to improving the performance of the coil gun. Keep a log book and take speed measurements after each change you make. Don't forget that if you change the coil, or the capacitors, you may find that the best starting position for the projectile may have changed. You will probably have to do tests at several different positions after each change.

In the next chapter, we stay with the weapons theme and build a trebuchet.

Trebuchet

PROJECT SIZE:	Small
SKILL LEVEL:	★★☆☆

SOMETIMES THE EVIL GENIUS LIKES to go for that retro-look when creating his projects, and it's hard to get much more retro than a medieval siege engine.

The trebuchet uses a weight attached to an arm with a sling on the end of it to throw objects. In its time, the projectiles varied, but were often big rocks or dead and rotting horses, thrown into the enemy fortifications during sieges. The Evil Genius' trebuchet will not throw much more than a tennis ball, but it will throw it a decent distance, and it's a good project to try if you like woodworking.

Figure 2-1 shows the trebuchet ready for action.

It's a simple design that should only take a few hours to construct and needs little in the way of special tools or equipment.

Trebuchets are elegant machines that convert the potential energy stored in a counter-weight into kinetic energy in the projectile. Unlike the coil gun of the previous chapter, trebuchets do this in an efficient manner. The action of a trebuchet is shown in the sequence of diagrams in Figure 2-2.

The trebuchet has a sling attached to the throwing arm. So, initially the projectile is almost

Figure 2-1   The Evil Genius trebuchet

underneath the weight (Figure 2-2A). As the weight falls, the throwing end of the arm rises, pulling the sling and projectile along the slide on the base of the trebuchet. At some point, the projectile leaves the slide (B) and is swung round in a wide arc as the weight keeps falling (C).

The sling has one end attached to the top of the throwing arm, and the other end attached to a ring that fits over a hook on the end of the throwing

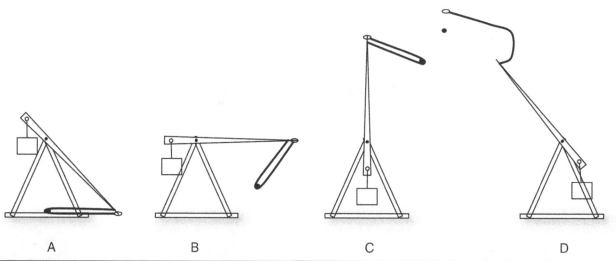

A   B   C   D

Figure 2-2	The action of a trebuchet

arm. As the arm passes the vertical mark, at some point the sling will slacken and the ring will slip off the hook, releasing the projectile (D).

WARNING!

Be careful when cocking the trebuchet, and stand well clear of it when it fires.

What You Will Need

You will need the materials listed in the Parts Bin to build this project.

All of these can be easily obtained from hardware stores, and in the case of the plastic container, a supermarket.

PARTS BIN		
Part	**Quantity**	**Description**
Lumber	5	1" × 2" × 8' (Furring strip) (25mm × 28mm × 2.4m)
Hardboard	1	Small sheet of hardboard to cut 24" × 9" (610mm × 230mm)
Metal plate	2	Flat metal plates 4" × 1.5" (100mm × 40mm)
Bolt	1	4" × ⅜" (100mm × 10mm)
Dry food container	1	10 pint (5L) plastic food container for containing breakfast cereal
Sand	25 lbs (10 kg)	Builder's sand or ballast
Rope	15 ft (5 m)	¼-inch (4mm) diameter nylon rope
Cloth	1	8" × 10" (200mm × 250mm)
Tennis balls	Variable	For use as projectiles
Metal hook	1	Metal cup-hook about 3 inches (75mm) when straightened out
Wood screws	Pack of 50	2-inch (50mm) wood screws
Wood screws	Pack of 10	3-inch (75mm) wood screws
Nail	1	6-inch (150mm) nail
String	1	3' (1m) string

Note that lumber is sold in different standard sizes in different countries, so you may not be able to get exactly the same size. This should not matter, and if in doubt, use thicker, more solid wood.

In addition, you will also need the following tools:

TOOLBOX
■ Wood saw
■ Drill and assorted drill bits
■ Assorted screws
■ Stapler or needle and thread

Assembly

The cutting list for this in Table 2-1 includes all the wood you will need for the project.

NOTE The author used inches in this project; it seemed appropriate for such an ancient design. So, those measurements are the more exact figures.

Step 1. Make the Frame Sides

The first step in this project is to build the frame sides. These are constructed from pieces of wood, which are arranged in an "A" shape. The apex of the "A" is held together by a metal plate, through which the bolt should be attached, which acts as a pivot. So, before attaching the plate, drill a hole right in the center, big enough for the bolt to pass through.

Figure 2-3 shows the plan for one of the A-frames.

One of the A-frames is shown in the photograph of Figure 2-4. The sections of wood are just screwed together with a pair of screws at each joint.

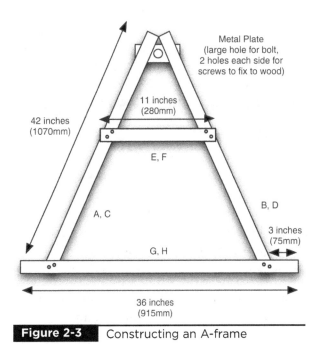

Figure 2-3   Constructing an A-frame

TABLE 2-1	Cutting List	
Piece	**Description**	**Dimensions**
A, B, C, D	A-frame sides	42 inches (1070mm) of lumber
E, F	A-frame bracing	11 inches (280mm) of lumber
G, H	A-frame base	36 inches (915mm) of lumber
I	Throwing arm	74 inches (1880mm) of lumber
J, K	Base ends	21 inches (530mm) of lumber
L	Center base, support for side struts	33 inches (840mm) of lumber
M	Center base, for trigger and runway	35 inches (890mm) of lumber
N	Runway	Hardboard
O, P	Side struts	41½ inches (1055mm) of lumber

Figure 2-4 An "A" frame

Step 2. Build the Throwing Arm

Figure 2-5 shows the construction of the throwing arm. Drill one large hole to fit the bolt used, and two smaller holes for the weight ropes and for attaching the permanent end of the sling.

Straighten out the metal hook (Figure 2-6) and screw it into the end of the throwing arm.

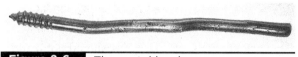

Figure 2-6 The metal hook

You will need to use pliers to grip the hook while you screw it into the end of the throwing arm.

Step 3. Build the Base

The structure of the base is shown in Figure 2-7.

We start by fixing the two A-frames together at the base, using two lengths of wood (J and K in the cutting list) placed under the A-frames. There should be a gap of about 7 inches (180mm) between the A-frames at the base.

Also attach the central strut (L) that will hold the side braces, and piece (M) that runs down the center of the base to support the runway board and trigger mechanism.

At the top of the frames, fit the bolt through one plate, a nut, and then the throwing arm and the second plate (Figure 2-7). Fit a nut on the inside of the bracket before the second plate. The nuts are going to hold the two A-frames apart against the side bracing.

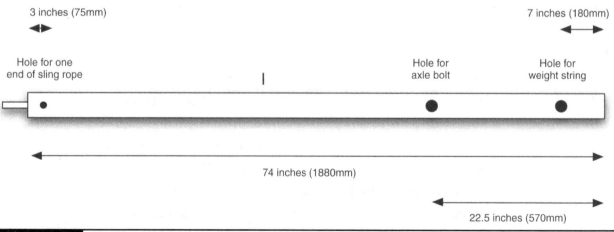

3 inches (75mm)

7 inches (180mm)

Hole for one end of sling rope

Hole for axle bolt

Hole for weight string

74 inches (1880mm)

22.5 inches (570mm)

Figure 2-5 The throwing arm

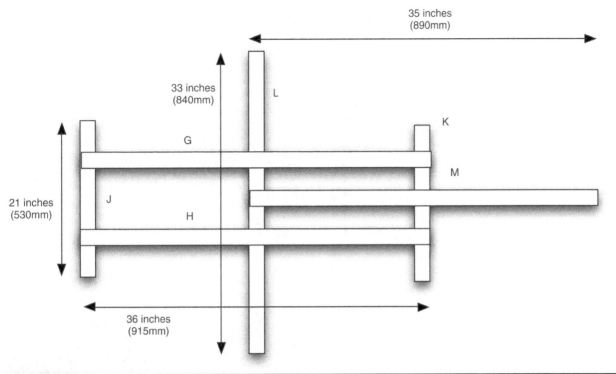

Figure 2-7   The base

Step 4. Attach the Side Braces

The side bracing is formed by a cross piece under the middle of the A-frames (L) and two struts (O

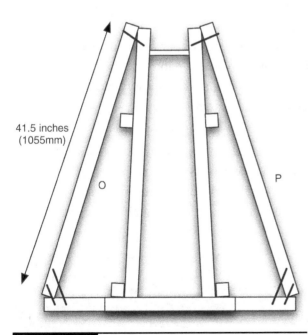

Figure 2-8   The base and side struts

and P) from the bottom of the strut up to the apex of the A-frames (Figure 2-8).

The struts and braces are both fixed into place using screws. Drill the struts at an angle first and use long screws. More adept woodworkers may elect to cut the ends of the struts at angles so they fit better.

Step 5. Assemble the Weight

The weight is constructed from a plastic box designed to hold breakfast cereal. The box has a 10-pint (5 L) capacity, which when filled with wet sand will have a weight of about 18 pounds (8 kg). Before filling the container, we need to drill four holes near the rim (Figure 2-9).

Cut two 24-inch (610mm) lengths of the nylon rope and thread them through the holes drilled in the food container and the hole at the weight end of the throwing arm. The rope should be just long enough to allow the weight to stay away from the rotating end of the throwing arm, without being so

Figure 2-9   The container used to make the weight

long that the weight hits the ground at its lowest position.

Check the travel of the whole mechanism, and make sure that there is enough clearance between the A-frames for the weight.

Figure 2-10 shows how the container will eventually be attached. Note that it is shown here filled with sand, but it is better to wait until everything is assembled before you fill the container.

Step 6. Assemble the Sling

The sling (Figure 2-11) is made from 64 inches (160mm) of the rope with a square patch of cloth,

Figure 2-10   The container in place on the trebuchet

8" × 10" (200mm × 250mm). A reasonably strong material like denim is ideal. The Evil Genius' minions can often be found wearing jeans with large patches of cloth removed. The Evil Genius tells them that this is the latest fashion and the minions are pleased.

Lay the rope across the cloth as shown in Figure 2-12 and fold the edges over to make a seam enclosing the rope. Sew the seams up, or if you prefer, apply a row of staples down each side. Sewing will be a lot more durable than staples.

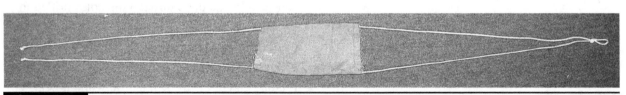

Figure 2-11   The sling

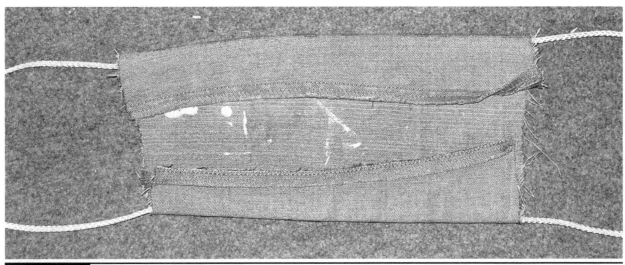

Figure 2-12 Making the sling pouch

Tie a loop into one end of the rope. This will fit over the hook of the throwing arm.

Step 7. Create the Runway

The runway is a rather grand name for the smooth panel of hardboard (N). It is along this that the projectile will be pulled before it is lifted by the rotating arm.

It fits on top of piece M (Figure 2-13) that sits across the base and doubles as the point to attach the trigger mechanism.

Step 8. Fashion the Trigger Mechanism

The trigger mechanism is quite unsophisticated. It comprises a nail on a string and a hole that goes through the throwing arm and the piece of wood M

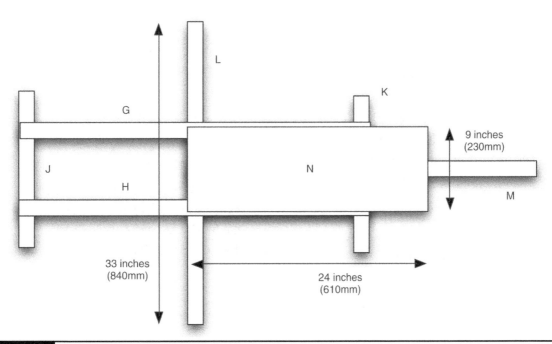

Figure 2-13 The runway

Figure 2-14   The trigger mechanism

(Figure 2-14). To fire the trebuchet, simply pull the nail out.

Fire!

A tennis ball makes a good projectile. It has the advantage of being fairly tough and not being damaged by a collision with a target (minion). To test the trebuchet, a suitable open space should be found and minions dispatched to a reasonable distance in front of the trebuchet.

Fit the ball into the sling and pull down the throwing arm until it can be pinned by the trigger nail. Stretch out the sling in the runway, with the ball at the far end, so that the rope is straight and the loop is over the pin on the end of the throwing arm.

Stand clear of the trebuchet and pull on the string to release the nail holding the throwing arm in place. The ball should sail off into the distance. Dispatch a minion to retrieve it and do it again, because it's fun.

Tuning the Trebuchet

There are various things you can do to the trebuchet to squeeze the best performance out of the engine.

You will likely need to make a few adjustments to the pin before this works well. If the ball flies off too low to the ground, then bend the pin back a little so the sling is released earlier. Or, if the ball is released too early and flies straight up into the air, bend the pin forward a little.

This book's web site (www.dangerouslymad .com) contains links to videos showing the trebuchet in action. These may be useful to refer to.

Theory

The trebuchet takes its energy from the weight that falls as the arm swings. This "potential" energy is transferred to the arm and sling of the trebuchet and is released as kinetic energy in the tennis ball. As we know the energy stored in the weight and know how far the tennis ball can be thrown, we can calculate both the energy going into the system and the energy released into the ball by the system. This will allow us to calculate the efficiency of our trebuchet.

The input energy can be calculated using the formula:

$$E = mgh$$

where m is the mass of the weight, g is the gravitational acceleration on Earth (9.8), and h is the height.

So the energy in joules is approximately:

$$E = 8 \text{ kg} * 9.8 * 0.5\text{m}$$

$$= 40 \text{ joules}$$

On the other hand, we can calculate the approximate amount of energy that was transferred into the ball on the basis of the distance that it traveled and its weight.

$$E = \frac{1}{2}mv^2$$

$$d = \frac{v^2}{g} \qquad v^2 = dg$$

So:

$$E = \frac{1}{2}mdg$$

The tennis ball weighs about 60 g and a good throw will send it 25m.

$$E = \frac{1}{2} * 0.06 * 25 * 9.8$$

$$= 7.35 \text{ joules}$$

From this, we can calculate the efficiency of our trebuchet at 7.35/40 = 18%.

This figure of 18% neglects the fact that the tennis ball has to travel through air and is probably not launched at the optimum angle. So, the real figure is probably closer to 25%.

This does not sound great, but actually is quite typical for a homemade trebuchet. These very rarely exceed 30%.

The following are a few things you can do to improve the efficiency of the trebuchet:

- Vary the length of the sling; start by making it longer.

- Decrease the weight of the throwing arm by reducing the amount of wood at the sling-end, or cutting it into a wedge shape.

- Vary the pin angle.

Just in terms of increasing the range, without changing the efficiency, increasing the weight is an easy win. You could use a bigger container, or use gym weights. You will, however, quickly need a stronger frame and throwing arm as you increase the weight.

Summary

If you want to see what happens when you scale up a design like this, take a look at YouTube. It shows some truly enormous trebuchets, which can throw people, pianos, barrels of flaming gasoline, even cars. Now that's evil!

In the next chapter, we will continue with the theme of weapons and make ourselves a ferocious minigun.

CHAPTER 3

Ping-Pong Ball Minigun

PROJECT SIZE: Small

SKILL LEVEL: ★☆☆☆

THERE ARE TIMES WHEN THE COIL GUN will not be powerful enough for the task that the Evil Genius has in mind. Sometimes you need to totally overwhelm your enemy with superior firepower. This minigun design will fire ten ping-pong balls per second at considerable speed. An entire magazine of 50 ping-pong balls can be emptied in just five seconds.

It is a very easy project to make and requires little in the way of tools or special equipment, and all the parts can be obtained from the Internet or your local hardware store (see Figure 3-1).

This project is seriously good fun and definitely more entertaining than blowing leaves about. However, it is strongly recommended that the Evil Genius have his minions on hand to pick up all the ping-pong balls, as this can become a bit tedious.

This design is described in such a way that you can adapt it for your particular leaf blower. Also, with the exception of a bit of glue, the project does not irreversibly alter the leaf blower, so it can

Figure 3-1 The leaf blower minigun

resume its original use should there suddenly be any leaves to blow.

WARNING!

This is not a real gun, and ping-pong balls will not generally damage humans unless shot at very close range, at the eyes, or swallowed.

However, a number of aspects of this project are dangerous.

- It uses electricity, and we do not modify the electrics of the leaf blower. However, you should observe the safety instructions that came with your leaf blower.
- Do not put your hands or fingers into the leaf blower when it is on or even just plugged in.
- Do not place your eye or anyone else's into the gun's line of fire.

What You Will Need

The components for this gun are all readily available. Leaf blowers are all slightly different, so you will probably have to modify this design to work with your particular leaf blower. Before buying the parts, however, read through the whole project and try out a few tests on your leaf blower to make sure it is suitable, and also to determine roughly what length of pipe you will need.

Obviously, the most important part of this project is the leaf blower. Various types of leaf blowers are available: electric ones, gas ones, ones that just blow, and ones that both blow and suck up leaves into an attached bag. The leaf blower used by the author is an electric one, which also has a bag for collecting the leaves. It is the aperture that the leaf collecting bag attaches to that is used to introduce the ping-pong balls.

If you have some other type of leaf blower, then as long as the air outlet is wide enough to accommodate a ping-pong ball and there is a spot in the air path where you can insert the balls, you should be able to make something. So, read through the rest of this chapter to understand the principals of what we are trying to do and then adapt them to work with your particular leaf blower.

The balls need to be a good fit with the pipe. They should fit into the pipe easily and run slowly through it if you tip it. If the balls are a lot smaller than the pipe, they will not be fired out as fast as they could be. On the other hand, if they cannot move freely enough through the pipe, they will become stuck.

That is why a 40mm internal pipe diameter and 38mm-diameter balls are a perfect combination. This ensures there is just enough space for our round projectiles.

PARTS BIN		
Part	**Description**	**Source**
	Electric leaf blower	Garage/hardware store
Barrels	1½-inch NPS 40 drain-pipe (internal diameter 40mm)	Hardware store Home Depot SKU: 193844
Magazine	Large plastic water bottle	Supermarket
Projectiles	50 × 38mm ping-pong balls	eBay
	Expanded polystyrene, enough to make two blocks that will fit into each end of the leaf blower	Waste packaging
	Duct tape	Hardware store

Be wary when buying your ping-pong balls, as the international standard size of balls since the year 2000 Olympic Games is now 40mm. Prior to that, 38mm balls were the most popular, especially in China. Plus, 38mm balls travel faster and can take more spin. This caused controversy at the time, as some saw the introduction of the bigger balls as a way of giving the non-Chinese competitors an advantage. This is clear evidence of the Evil Genius at work in the world of table tennis.

Plenty of people still sell 38mm balls (often from China), but check before you buy. You will also find "plastic" ping-pong balls on eBay. These are made of soft plastic, and tend to be made to far looser tolerances than proper celluloid balls. So even if these types of balls are advertised as 38mm, they may not be exactly 38mm, so try and avoid these, unless you already have some samples to try in a pipe.

It is also a good idea to take a ball along with you when you go to buy the pipe.

In addition to the parts just listed, you will also need the following tools:

T O O L B O X
■ Hacksaw
■ Scissors/craft knife
■ Epoxy resin glue or hot glue gun

Assembly

The Evil Genius still has a yard in which leaves must be blown, so as we mentioned earlier, this project does allow his leaf blower to revert to its original use in the fall. Better yet, it is an easy project to construct, since it is a simple design.

The project adds three barrels to the leaf blower that are made of plastic wastewater pipe. These help channel the ping-pong balls, increasing the accuracy of the gun. There is no complex magazine or trigger mechanism; instead, the magazine is hinged onto the side of the leaf blower and the balls are simply tipped into the leaf blower.

When used as a gun, the leaf blower is held upside down to allow balls to be poured into the aperture where leaves are normally collected.

The leaf blower used by the author is shown in Figures 3-2, 3-3, and 3-4. Unless you have the same model, you will probably find that your leaf

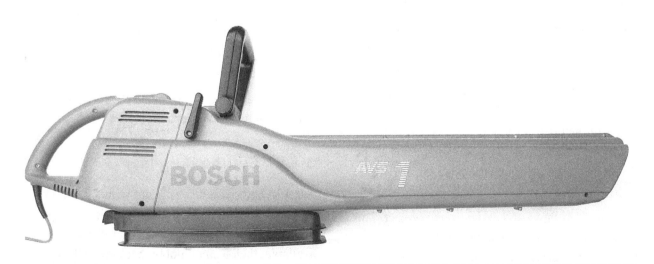

Figure 3-2 Leaf blower, side view

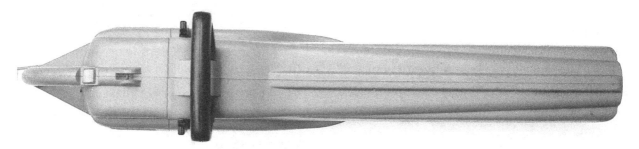

Figure 3-3   Leaf blower, top view

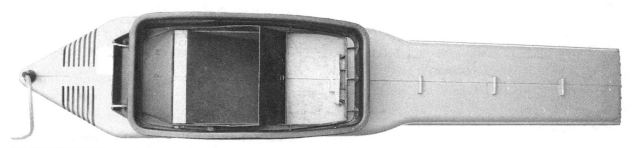

Figure 3-4   Leaf blower, bottom view

blower is a different shape, but with a very similar basic design.

Familiarize yourself with the design of your leaf blower and compare it with the author's. Turn it on, and if it has different settings, find the setting for maximum blowing power out of the nozzle. Also, look through it to see the path the air takes. You can also try dropping a ping-pong ball into it at the air-intake while it is running. Do *not* put your fingers inside the blower. This should be fired out, albeit not very fast or accurately. Work out where the balls will need to be introduced, as this will determine where you attach the magazine.

Your leaf blower may have a round rather than rectangular nozzle. If this is the case, you can either fit the three waste pipe barrels in a row, as described in these instructions (and perhaps adapt things by fitting them in a triangular arrangement), or use just two barrels.

Figure 3-5 shows the design of the project. This is a simple design; the leaf blower does not need a lot of modification. Essentially, it just needs barrels to direct the balls accurately and a container for the balls. The container is hinged so that to fire the gun, balls are simply tipped into the leaf blower.

As we mentioned earlier, the leaf blower is actually held upside down to allow the balls to be tipped into the inlet, where they fall into the path of the air jet, which whisks them into the barrels and fires them out the other end.

The barrels are wedged into place with blocks of expanded polystyrene packing material that has been cut to the right size. This also has the effect of blocking most of the air that would otherwise flow around the pipes, thus maximizing the flow through the pipes.

The following sections take you through the construction step by step.

Step 1. Cut the Barrels

The PVC pipe can be easily cut with a hacksaw. You will get a cleaner, more perpendicular edge if you use a large sturdy hacksaw, but if none is available, then a small hobby hacksaw will work fine. They should be cut to a length where one end

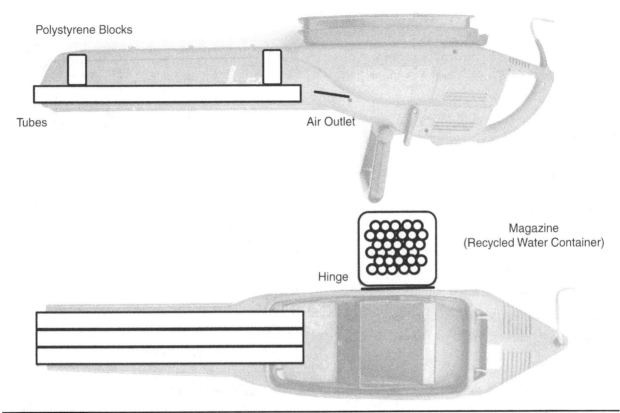

Polystyrene Blocks

Tubes

Air Outlet

Magazine
(Recycled Water Container)

Hinge

Figure 3-5 The minigun design

of the barrel protrudes from the nozzle of the leaf blower by about an inch (25mm) and the other end stops about three inches from the air outlet of the blower (see Figure 3-5).

This should allow the maximum flow of air into the pipes. Having the pipes protrude from the end of the leaf blower serves to make the device look more gun-like. Without it, the enemy of the Evil Genius might get the misleading impression that they are simply about to be blown to death and fail to take the Evil Genius seriously. That is something no Evil Genius will tolerate.

The PVC pipe is surprisingly easy to cut. When cutting, first draw a line around the pipe where you want to make the cut. Then, unless you have a large sturdy hacksaw and vice, cut around the line, shifting position as you start to cut into the middle. Once cut, use a file or knife to scrape around the newly cut rim and remove any burrs. You should then be able to put a ball in one end and have it freely roll to the other and fall out.

Having cut the barrels, tape them together with duct tape a few inches back from each end, as shown in Figure 3-6. This combines them into a single solid structure. As mentioned earlier, if the pipes will not fit into the leaf blower side by side, you can experiment with either taping them together in a triangular pattern, or just using two barrels.

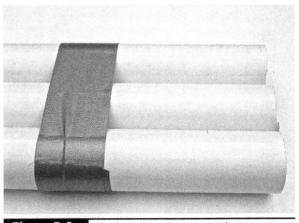

Figure 3-6 Taping the pipes together

Before you fix anything in place, hold the pipes where they leave the leaf blower, turn the leaf blower on, and find the position for the pipes where you get the strongest blast of air through them.

Step 2. Fit the Barrels

The pipes are held in place by blocks of expanded polystyrene packing material (Figure 3-7). These are slightly larger than the gap they are filling and wedge the pipes firmly in place. They do not provide a complete seal for the pipes, but they do block most of the air flow around the pipes, so nearly all the air is directed into the pipe.

Figure 3-8 shows the block for the nozzle end, and Figure 3-9, the one for the outlet end. Notice

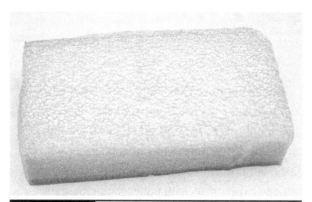

Figure 3-7   Blocks of expanded polystyrene packing material

Figure 3-8   Pipes in place at the nozzle end

Figure 3-9   Pipes in place at the air inlet end

in Figure 3-9 how the pipes line up with the air outlet from the outlet vent at the bottom of the figure.

Before we go any further, let's do a quick test of our gun. First of all, select a disposable minion and stand them about 20 ft (9m) from the gun. Now set the leaf blower to blow mode, turn it on, and pop a ball in front of one of the pipes near the air outlet. This should fire the ball at a considerable speed at the minion. Repeat the experiment, throwing a handful of balls into the "firing chamber" and watch the minion dance for your amusement.

Picking a minion that wears glasses will reduce the chances of damaging their eyes.

Step 3. Create the Magazine

A large plastic water bottle was used to hold the balls. The top of this, where it narrows to the cap, is cut in such a way as to leave a flap that can be bent into a hinge (Figure 3-10), producing a shape that is basically cuboid. The hinged flap is then glued to the side of the leaf blower.

To fire the gun, simply tip some balls from the magazine into the leaf blower inlet. After much experimentation with various trigger mechanisms, this was found to be the most effective way of controlling the firing of the gun.

Figure 3-10   The magazine attached to the leaf blower

Testing

It is now time to try out the gun.

The gun should be used indoors, or you are likely to lose the ping-pong balls. Once again, a minion is required as a target. Stand at least five yards (meters) away from the minion. After giving them strict orders not to step on any of the ping-pong balls, fire up the leaf blower and tip in a few balls to get your aim right. Once your aim is accurate, pour in the rest of the balls as quickly as possible.

Finding the Muzzle Velocity

We can measure our muzzle velocity using the same approach we did with the coil gun (see Chapter 1). We will only need a single ping-pong ball and the impact target must be placed much further away. The arrangement we used is shown in Figure 3-11.

As a target for measuring the speed of the balls, a wall, window, or other solid object is much better than a minion. Minions are soft, and while they do make a noise when hit by a ball, a considerable delay occurs as the message slowly propagates through their nervous system and eventually reaches their brain. This delay tends to spoil the accuracy of the figures.

Your laptop should be positioned close to the target rather than the leaf blower to reduce the extraneous noise created by the leaf blower.

The leaf blower is very noisy, so it is much harder to tell from our sound trace exactly where the ball left the nozzle of the gun. A trace is shown in Figure 3-12. The range between when the ball leaves the gun and when it hits the test target is marked on the diagram. The only way to determine the moment the ball leaves the gun is to repeatedly play part of the sound wave. Your built-in signal processor (brain) will be able to separate the "pop"

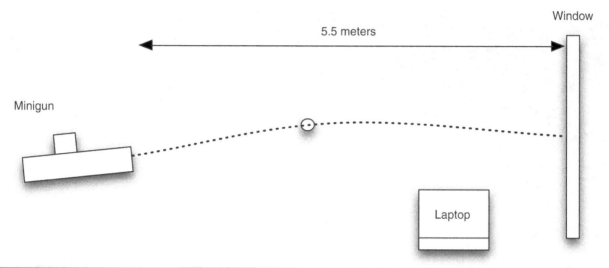

Figure 3-11 Finding the muzzle velocity

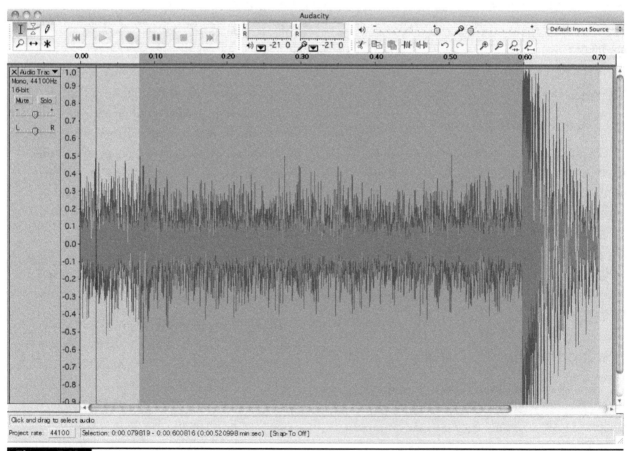

Figure 3-12 The sound waveform of a firing

as it leaves the nozzle from the general roar of the leaf blower.

From this trace, we can see that our ball took 0.6 – 0.08 or 0.52 seconds to travel 5.5 meters, indicating an average velocity of 10.6m/s, or 24 miles per hour. Being very light and having a large surface area, the ping-pong ball will lose speed quickly, so the muzzle velocity is likely more than this figure.

An alternative to measuring the velocity with sound would be to use video. Use a similar arrangement to the sound, but set up a video camera at right angles to the flight of the balls. If you know the number of frames per second used by the video camera, then by estimating the distance the ball has moved between frames, you can work out the velocity.

The only problem with this approach is that you need a lot of room so you can put the video camera far enough away from, and to the side of, the flight of the balls. The author did not have a lot of success with this approach; the balls just moved too fast. But you may have more room for such experiments and a faster camera.

Theory

A leaf blower is a low-pressure air pump. It is designed to blow out as much air as possible, as fast as possible. By reducing the effective area of the end of the nozzle by fitting the barrels over just half the area, we can increase the velocity of the air traveling through the barrels.

Since the leaf blower is open at both ends, if we increase the resistance to flow too much—say, by just having one barrel and blocking up the rest of the tube—then the air may simply bounce off the obstruction and come out the back of the blower, or create a vortex that merely circulates the air and hence the balls.

Trial and error was used to determine the best number of barrels to maximize the speed of the balls and the fire rate. For the author's leaf blower, this turned out to be three barrels, which conveniently fit across the width of the blower.

Summary

This is one of the most enjoyable and easily completed projects in the book. As you would expect, applying a few kW of power to a bunch of ping-pong balls is always going to produce some exciting results.

In the next chapter, we will embark on another weapon-type project, but this time it will be a model weapon, based on a laser and some servo motors.

CHAPTER 4

Mini Laser Turret

PROJECT SIZE: Medium

SKILL LEVEL: ★★★★

WHEN PLANNING A CAMPAIGN of world domination, the Evil Genius likes to set up a model battlefield using small plastic warriors on a realistic but miniature battleground. Such preparations are made even more enjoyable if one of the Evil Genius' favorite toys is brought into play.

The mini laser turret (Figure 4-1) can sit amidst the miniature battlefield. A joystick allows the laser to be directed in a sweeping arc across the battlefield, cutting a deep swath through the massed ranks of the enemy.

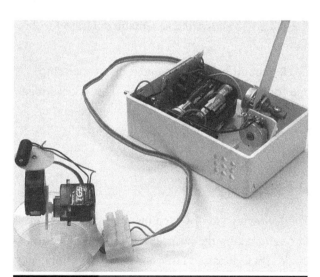

Figure 4-1 Mini laser turret

The turret gun is made of two miniature servos, of the type used in remote control planes, mounted at right angles to each other so that one sweeps left to right and the other raises or lowers the laser module. Wires connect the turret to a control unit with a homemade joystick that allows the laser to be aimed remotely.

This project can also be combined with the project in the next chapter to allow the popping of balloons or the jumping of a can by the laser.

WARNING!

You should always take certain precautions when using lasers:

- Never shine a laser into your eyes or anyone else's.

- Resist the temptation to check if the laser is on by peering into it. Always shine it onto a sheet of paper or some other light surface instead.

What You Will Need

To build this project, you will need the components shown in the Parts Bin on the next page.

You can buy laser modules from standard component suppliers like Farnell and RS, but they tend to be very expensive. So look online, where they should cost no more than two or three dollars. The same applies to the servo motors, and if you

PARTS BIN			
Part	**Quantity**	**Description**	**Source**
Laser	1	3 mW red laser LED module	eBay
IC 1	1	NE556 dual timer IC	Farnell: 1094320
R1, R7	2	1kΩ variable resistor (linear potentiometer)	Farnell: 1174082
R2, R4, R6, R9	4	1kΩ 0.5W metal film resistor	Farnell: 9339779
R3, R5, R8, R10	4	10kΩ 0.5W metal film resistor	Farnell: 9339787
R11	1	100Ω .05W metal film resistor	Farnell: 9339760
T1, T2	2	BC548	Farnell: 1467872
C1, C3	2	1µF 16V electrolytic capacitor	Farnell: 1236655
C2, C4	2	10nF ceramic capacitor	Farnell: 1694335
Servos	2	9 g servo motor 4.5–6V	eBay
Battery clip	1	PP3 battery clip	Farnell: 1650667
Battery holder	1	Holder for 4 AA batteries	Farnell: 1696782
IC socket (optional)	1	14-pin DIL IC socket	Farnell: 1101346
S1	1	SPST toggle switch	Farnell: 1661841
Box	1	Small plastic project box	
Terminals	1	Four-way terminal block	Farnell: 1055837
Stripboard	1	Stripboard, 15 tracks each of 20 holes	Farnell: 1201473
Cable		Four-way ribbon cable, or shielded four-way cable for longer distance control (1 meter)	Farnell: 150431
Base	1	Aerosol can lid	Recycling

are happy to wait a week or two for your parts to come from China, there are some bargains to be had this way, too.

In addition, you will also need the following tools:

TOOLBOX
■ An electric drill and assorted drill bits
■ Soldering equipment
■ Epoxy resin glue or a hot glue gun
■ A craft knife

Figure 4-2 shows the schematic diagram for this project.

The circuit is based around a dual timer chip, one for each of the servo motors. This is described more fully in the "Theory" section at the end of this chapter.

Assembly

The following step-by-step instructions walk you through making the laser turret. First, we construct the joystick, then the laser and servo assembly, and then the electronics that link it all together.

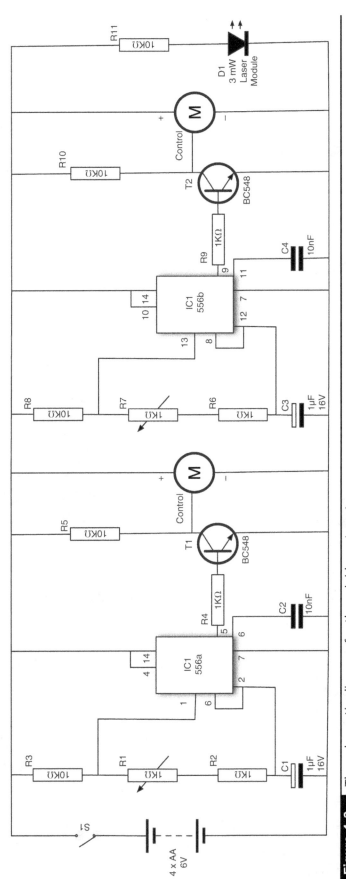

Figure 4-2 The schematic diagram for the mini laser turret

Step 1. Make a Frame for the Joystick

The construction of the joystick is shown in Figure 4-3. The two variable resistors are fixed to each other at the flat part of the stems by drilling a small hole through them and then fixing a nut and bolt through the two. The handle is cut from the same plastic, a hole drilled in it, and then mounted on one of the variable resistors.

The frame for the joystick is made from a right-angle plastic molding, cut to shape with scissors (Figure 4-4). Any suitable plastic of reasonable thickness is fine for this. The handle for the joystick is made from the same material (Figure 4-5). Both pieces are drilled to fit the variable resistor.

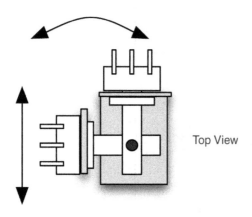

Top View

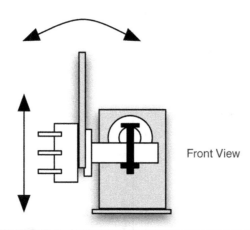

Front View

Figure 4-3   Construction of the joystick

Figure 4-4   The joystick frame

Figure 4-5   The joystick handle

Step 2. Fix the Variable Resistors Together

After cutting and drilling the plastic for the frame and handle, fit the variable resistors and then join them together through the drilled holes (Figure 4-6).

Step 3. Finish the Joystick

Solder 4-inch (100mm) leads to the center and right connections (looking from the back of the variable resistor) of each variable resistor. These will be attached to the stripboard when we have completed it.

Figure 4-6   The variable resistors attached to each other

Step 4. Fix the Laser and Servos

Figure 4-7 shows the design and wiring of the turret module. The laser module is mounted onto one of the servos. This servo is then mounted onto the arm of the other servo so the bottom servo will control the vertical angle of the laser, and the top servo the horizontal angle.

The servos are usually supplied with a range of "arms" that push onto a cogged drive and are secured by a retaining screw. One of the servos is glued onto one of these "arms" (Figure 4-8). Then, the arm is attached to the servo. Do not fit the retaining screw yet, since you will need to adjust the angle. Glue the laser diode to a second "arm" and attach that to the servo. It is a good idea to fix some of the wire from the laser to the arm in order to prevent any strain on the wire where it emerges from the laser. You can do this by putting a loop of solid core wire through two holes in the server arm and twisting it around the lead (again, see Figure 4-8).

Next, cut a slot in the can lid so the servo arm can move (Figure 4-8). Afterward, glue the bottom servo to the lid. Make sure you understand how the servo will move before you glue it.

Once you are sure everything is in the right place, fit the retaining screws onto the servo arms. You may need to adjust these once you come to test the project.

Step 5. Wire the Servos and Laser

The wires from the servos and the laser module will all be connected up to a terminal block (Figure 4-9). You will need to shorten the leads, and it is much easier to fit them into the terminal blocks if you solder the leads together first.

The colors of the leads on the servo vary among manufacturers. The leads on the author's servos were brown for GND, red for +V, and orange for the control signal. Check the datasheet for your servos to make sure you have the right leads.

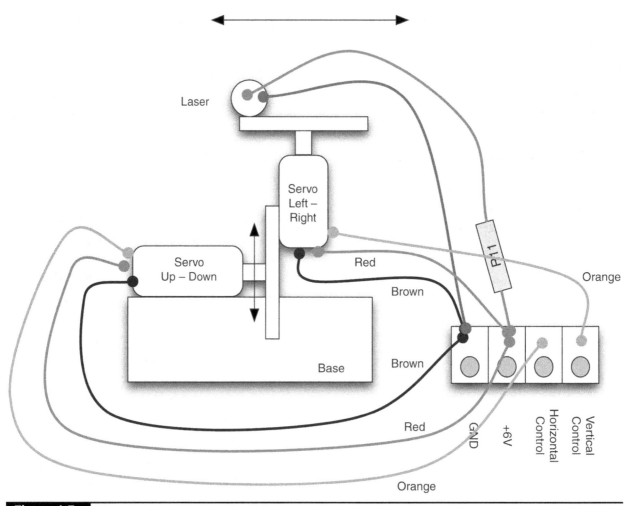

Figure 4-7 The design of the turret module

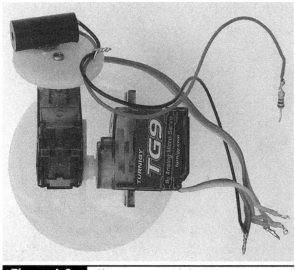

Figure 4-8 The servo module

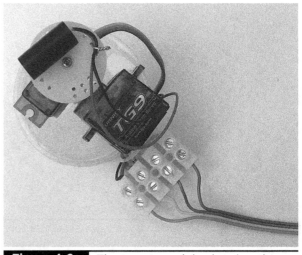

Figure 4-9 The turret module showing the wiring

Using Figure 4-7 as a reference, wire up the turret.

All the negative leads from both servos and the laser module go into the left-hand terminal block. The next terminal block has the positive leads of

the servos and the 100Ω resistor. The resistor leads should be shortened and the positive lead of the laser module connected to the end of the resistor.

The final two connections for the terminal block are the control signals from the servos.

Figure 4-10 shows the completed laser turret module, with the ribbon cable attached and ready to be connected to the stripboard.

Step 6. Prepare the Stripboard

Figure 4-11 shows the stripboard layout for the project. Note that it is shown as viewed from above.

Begin by cutting the stripboard to size: we need 15 strips, each with 20 holes. A strong pair of scissors can do this just fine. Using a drill bit, and twisting it between your fingers, cut the track in the locations marked with an "X." Figure 4-12 shows the back of the board, ready for soldering.

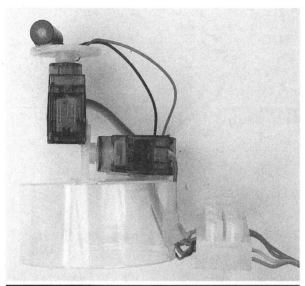

Figure 4-10 The completed turret module

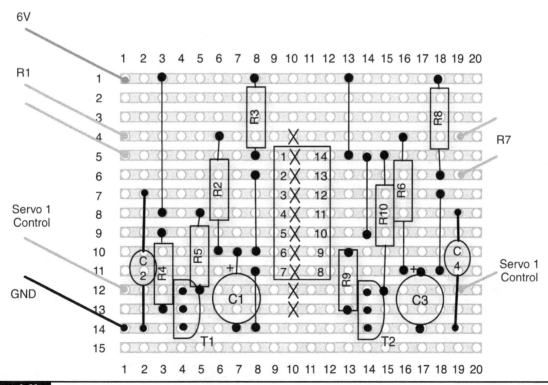

Figure 4-11 The stripboard layout

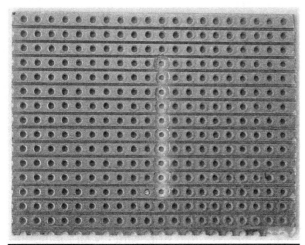

Figure 4-12 The prepared stripboard

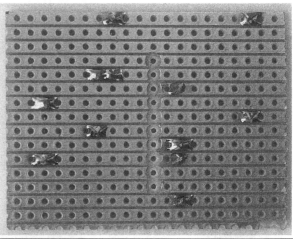

Figure 4-14 The bottom of the stripboard with the linking wires in place

Step 7. Solder the Components

First solder in the linking wires. Use solid core wire, either by stripping normal insulated wire, or by using previously snipped component leads. When all the linking wires are in place, the top of the board should look like Figure 4-13, and the bottom of the board like Figure 4-14.

The trick with the stripboard is to solder the components that rise least from the board first. That way, when you put the board on its back, they are held in place by the board while you solder them. This being the case, we are going to solder the resistors next.

Figure 4-15 shows the board with the resistors in place.

We can now solder the timer chip. You may choose to use an IC socket rather than solder the chip directly onto the board. If you do decide to solder the chip directly, be very careful to put it the right way around and in the right place. Once soldered into place, it is very hard to remove the chip. Also, be careful not to overheat the chip while soldering. Try to do it quickly, pausing a few seconds after soldering each pin.

After the chip (or socket), solder the small capacitors (C2 and C4) and the transistor. Again, check that the transistors are the right way around.

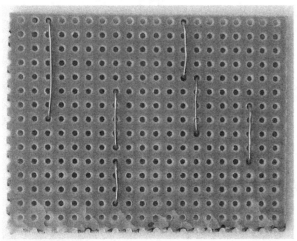

Figure 4-13 The top of the stripboard with the linking wires in place

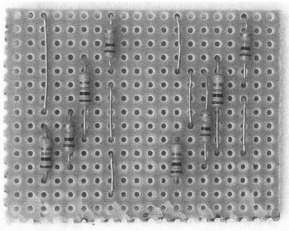

Figure 4-15 The top of the stripboard with the resistors in place

Finally, we can solder the large capacitors (C1 and C3) into place, making sure the polarity is correct. The negative lead is the shorter of the two leads and often has a diamond next to it.

The completed board is shown in Figures 4-16 and 4-17.

Figure 4-16   The completed stripboard from the top

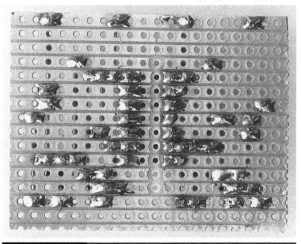

Figure 4-17   The completed stripboard from the bottom

Step 8. Wire Everything Together

Having built the turret module joystick and stripboard, it is now time to wire everything together. Figure 4-18 shows the wiring diagram of the project.

Using Figure 4-18 as a guide, wire together all the components. The ribbon cable to the servo module is fine for a length of a few feet, but you may run into trouble if you try and attach a long cable. Eventually, the pulse width signal will become distorted and the servos will behave unpredictably. Alternatively, to get a longer range with your wires, use multi-core shielded cable, where each of the control signals is in its own screened cable.

So, now that everything is connected, perform a final check to make sure there are no bridges of solder on the stripboard and that all the wires are in the right place. Afterward, insert the batteries and turn it on. The servos should "snap" to the position of the joystick.

If nothing happens, or only one of the servos works, disconnect immediately and check everything over.

Step 9. Adjust the Servo Arms

At this stage, you will probably need to adjust the position of the servo arms.

With the project turned on, set the joystick to its center position. Remove the two server arms and refit them so the servo is pointing the laser straight ahead, both horizontally and vertically.

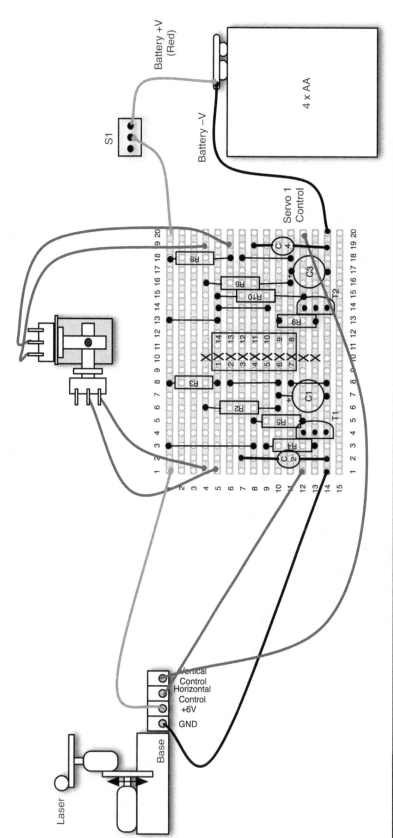

Figure 4-18 The wiring diagram

Step 10. Fit the Project into a Box

Because the joystick has quite a wide range, this is a difficult project to box. However, fitting the battery, the stripboard, switch, and joystick into a box without a lid will at least neaten up the project and provide a firm base for the joystick (Figure 4-19).

A hole is drilled for the switch, and another for the cable to the servo turret. The joystick frame is glued to one side of the box. The adventurous Evil Genius may even decide to provide a lid for the box.

Ideas

The Evil Genius might like to make more than one turret connected to the same joystick. That way, when the joystick is moved, all the turrets move. This is very easy to do. Each turret servo assembly is essentially connected in parallel, as shown Figure 4-20.

Theory

When it comes to making things move, servo motors are great. They are easy to use, quite low power, and only require one wire to control their position. In this section we learn a little more about how to use Servo motors.

We also take a look at the 555 timer IC, which is used to generate the pulses for the project. This timer chip is extremely versatile and it is useful to know how to use it.

Servo Motors

Servo motors are most commonly used in radio-controlled model vehicles.

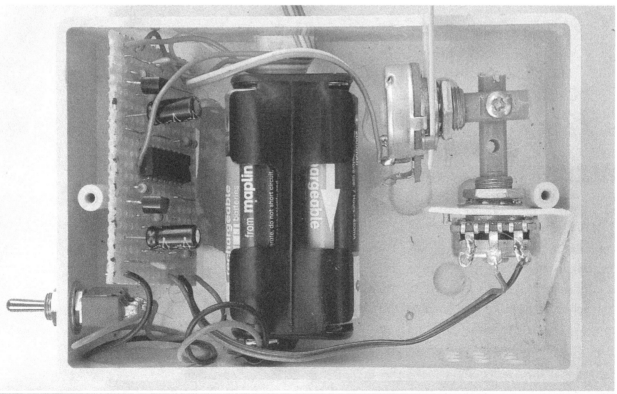

Figure 4-19   Boxing the project

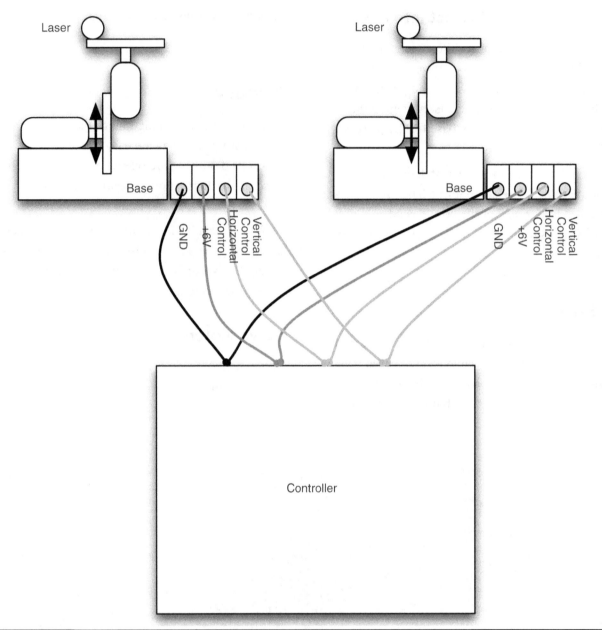

Figure 4-20   Controlling more than one turret from a joystick

Unlike standard motors, a servo module does not rotate around and around. It can only travel through about 180 degrees. The modules are controlled by pulses of voltage on the control connection to the servo. The length of the pulse controls the angle to which the servo is set.

Figure 4-21 shows an example waveform. You can see how the width of the pulses varies the servo angle.

A pulse of 1.5 milliseconds sets the servo to its center position, a shorter pulse of 1.25 milliseconds to its leftmost position, and 1.75 to its rightmost position.

The servo motor expects there to be a pulse at least every 20 milliseconds for the servo to hold its position.

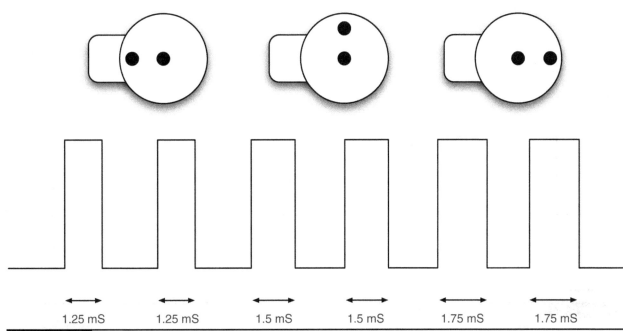

| 1.25 mS | 1.25 mS | 1.5 mS | 1.5 mS | 1.75 mS | 1.75 mS |

Figure 4-21 Servo motors

The 555 and 556 Timer ICs

To generate the pulses for our servo, we use a timer chip. Actually, the chip contains two timers in one package. That is, one for each servo.

The chip is an NE556, which contains the equivalent of two NE555 chips, one of the best-selling chips ever created. This chip has been around since 1971 and sells around a billion units a year. It can be used as both a one-shot monostable that produces just a single pulse, or as an astable oscillator that produces a stream of pulses.

We use it in this astable mode. The overall frequency and duration of the pulses are controlled by the timing components, comprised of a capacitor and two resistors. Figure 4-22 shows the basic arrangement on which our design is based.

Using this arrangement, as we said before, the circuit will oscillate, producing a waveform similar to that shown in Figure 4-23.

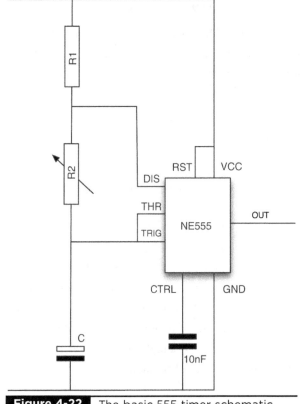

Figure 4-22 The basic 555 timer schematic

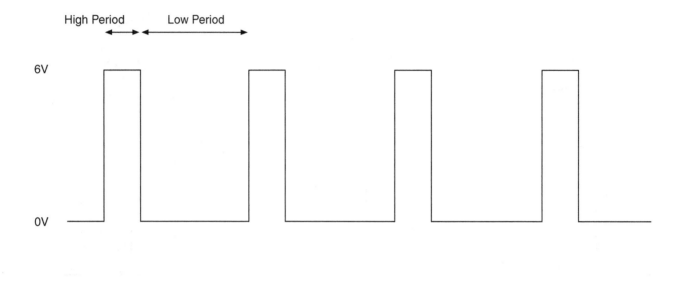

Figure 4-23 The 555 timer waveform

The "high period" shown in Figure 4-23 can be calculated using the formula:

$$0.69 * (R1 + R2) * C$$

where R is in Ω, and C is in F. A 1F capacitor is actually a huge capacitance. Capacitance values are usually measured in μF (microfarads) or even smaller units.

The "low period" shown in Figure 4-23 uses the formula:

$$0.69 * R2 * C$$

Notice that the low period only depends on the value of R2; R1 has no effect on it. This means that we can change the value of R2 to vary the length of the pulse. In our circuit, we use a combination of a variable resistor and a fixed resistor as an equivalent to R2, the total resistance being the sum of the variable resistor and the fixed resistor. The overall frequency will vary, but for servo motors, it's the length of the pulses that matter, not the overall frequency.

This system has one snag: We can easily vary the low period, but the servo motor is controlled by the high period. This is why we use a transistor connected to the output signal. This transistor inverts the signal, converting the high into low, and vice versa.

Summary

In this project, we learned how to make a joystick from a pair of variable resistors, as well as how to use servo motors. We also explored the extremely versatile 555 timer IC.

As an alternative to controlling the servos using a joystick, we could also control them with an Arduino interface board. We will use these small microcontroller boards in Chapters 8, 13, and 15. If you are interested in computerizing your servos and laser, there is a design in the book *30 Arduino Projects for the Evil Genius* by this author (Simon Monk) that does exactly that.

CHAPTER 5

Balloon-Popping Laser Gun

PROJECT SIZE:	Large
SKILL LEVEL:	★★★★

CONTINUING WITH THE GUN THEME, this project will allow the Evil Genius to shoot tin cans off a wall and burst balloons with a tiny but super-powerful laser gun, essentially your stock sci-fi "ray gun" (see Figure 5-1).

Well, actually, as you might have come to expect from the Evil Genius, there is a bit of trickery going on here.

The gun itself is just a low-power laser, and if you wish, you can simply use a standard laser pointer. The clever bits are in the "can" or the "balloon popper." Both of these have a sensor that detects the laser light and an electronic circuit that turns on a high-power transistor.

In the case of the "can," this transistor powers an electric motor with a weight attached. The weight is swung around and hits the inside of the can causing it to jump.

The balloon popper is very similar, but this time the transistor passes a large current through a small resistor, heating it up until it is hot enough to pop the balloon attached to it. The power flowing through the resistor is likely to eventually destroy it. Fortunately, the resistors are cheap and for this project can be considered disposable.

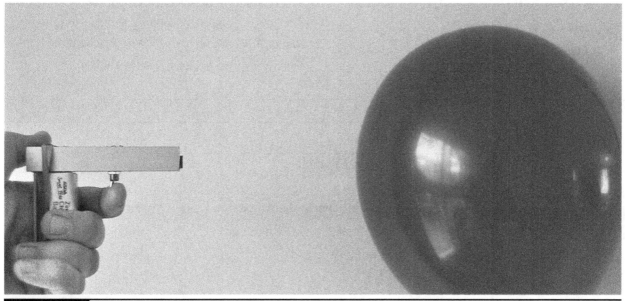

Figure 5-1 The ray gun in action

WARNING!

Though this is not a real gun, and is actually a low-power laser, you should still be aware of certain things:

- Never shine a laser into your eyes or anyone else's.

- Resist the temptation to check if the laser is on by peering into it. Always shine it onto a sheet of paper or some other light surface.

- The balloon-popper resistor will get very hot. Plenty hot enough to burn you, so do not touch it while it is on.

- Some resistor types can produce flame when used in this way, so keep the balloon popper far from anything flammable.

Assembly

The project construction is split into three parts, with three separate parts lists and sets of instructions:

- The ray gun

- The balloon popper

- The can jumper

The Ray Gun

For the "ray gun," you can just use a laser pen, or build a more gun-like module using the following components.

What You Will Need

You can buy laser modules from standard component suppliers like Farnell and RS, but they tend to be very expensive. So look online, where they should cost no more than two or three dollars. You'll need the parts shown in the Parts Bin.

In addition, you will also need the following tools:

TOOLBOX

- An electric drill and assorted drill bits
- A metal file
- Epoxy resin glue or a hot glue gun
- Soldering equipment

Assembling the Gun

The first thing to say about the gun is that you don't actually have to make it. You could just use a ready-made laser pointer. Laser pointers do not look very gun-like, so as an alternative to this design you could modify a pointer by adding a handle and trigger.

The gun (Figure 5-2) is deliberately made as small and unimpressive as possible to increase the impact of its balloon-popping and can-jumping capabilities. Figure 5-3 shows the schematic diagrams for the gun. It has only a few

PARTS BIN

Part	Quantity	Description	Source
Chassis	4 inches (200mm)	u-section aluminum	Hardware store
S1	1	SP toggle switch momentary on	Farnell:1550179
D1	1	3-mW red laser LED module	eBay
R1	1	100Ω 0.5 W	Farnell: 9339760
Battery clip	1	PP3 battery clip	Farnell: 1183124
Battery	1	PP3 9V battery	

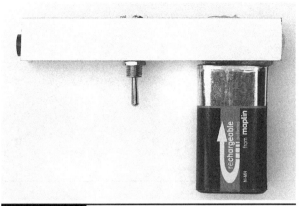

Figure 5-2 The ray gun

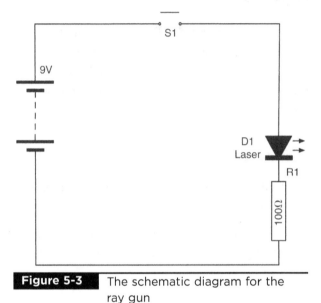

Figure 5-3 The schematic diagram for the ray gun

from hardware stores. The components could just as easily have been used on a strip of plastic or even built into a toy gun.

The battery clip terminals fit through holes made in the aluminum so that when a PP3 battery is attached, it forms the handle of the gun.

The resistor limits the current to the laser diode module. Resist the temptation to buy just a laser diode; instead, look for a laser diode module. The difference is that the "laser diode" will not have a lens, so you will not get the tight beam of a laser.

Step 1. Drill the Aluminum Chassis

The u-section aluminum that the author used is shown drilled and filed in Figure 5-4. Be careful to make the holes for the battery clip big enough to ensure that they will not make contact with the aluminum and cause a short circuit.

Lay out the components as they will fit onto the chassis and mark with a pencil where you need to drill. You will need a drill bit that's the correct diameter for the toggle switch and a larger bit to make the holes for the battery clip.

After drilling, file off any burrs in the aluminum and file the holes drilled for the battery clip into a square so the clips can fit in place without the contacts touching the aluminum. Try it on for size (Figure 5-5).

components and they can easily be soldered together without the need for a circuit board.

The components are mounted inside a u-section aluminum strip. Lengths of this can be obtained

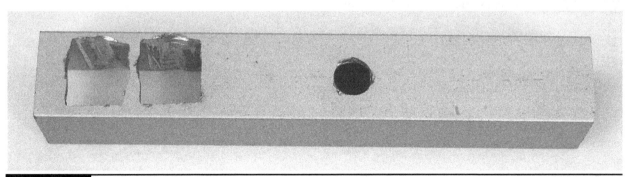

Figure 5-4 Drilling the chassis

Figure 5-5   Checking that the battery clip fits

Step 2. Solder Everything Up

It is easiest to solder together all the components before fitting them into place. It also means that you can check that it works okay before you glue everything down.

Using the wiring diagram of Figure 5-6 as a guide, shorten the leads of the laser module, battery clip, and resistor to the right lengths. Strip the ends of the insulated wires and solder the components together as per the wiring diagram.

After everything is connected up, try operating the switch to make sure the laser lights up before moving on to the next step.

Step 3. Final Assembly

Fit the retaining screw over the switch and tighten it with pliers. Then, using either epoxy resin glue or a hot glue gun, glue the laser module and battery clip into place. Figure 5-7 shows the parts soldered and glued into place.

Testing the Ray Gun

The Evil Genius likes to test laser guns with the aid of a pet.

Shine the laser in front of your pet (avoid the eyes) and watch them try to catch the red spot on the floor or on the walls.

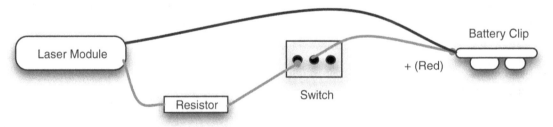

Figure 5-6   The wiring diagram for the ray gun

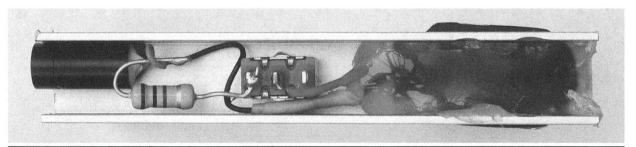

Figure 5-7   The assembled ray gun from above

Cats are much better for testing than dogs, because most dogs will abandon the chase once they have worked out that they cannot eat the little red dot. Cats, on the other hand, will happily chase the elusive dot around the room for some considerable time. In this respect, they are somewhat similar to minions.

The Balloon Popper

In this section, we will describe how to make the balloon-popping part of the project. If you are more interested in shooting cans than in popping balloons, skip this section.

The balloon popper (Figure 5-8) is a small box containing a battery, a light sensor, and some other electronics. When light from the ray gun hits the light sensor, it turns on a transistor that allows a large current to flow through a resistor. The resistor is fixed to a terminal block and taped to the balloon. The resistor gets hot enough to burst the balloon after 10 or 15 seconds.

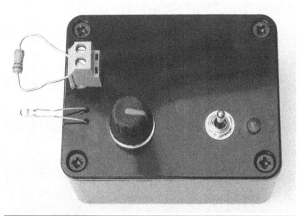

Figure 5-8 The balloon popper

The balloon popper has a control knob that sets the sensitivity of the sensor, and a switch that can be set to "live," "off," or "test." When set to "test," the resistor does not get hot; instead, an LED lights up. This lets you set the correct sensitivity.

The electronics for the balloon popper are contained in a plastic box. Most of the components are mounted onto the lid of the box. Figure 5-9 shows the schematic diagram for the balloon popper.

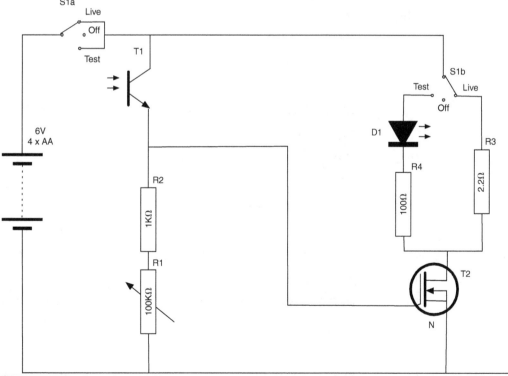

Figure 5-9 The schematic diagram for the balloon popper

What You Will Need

You will need the components listed in the Parts Bin for the balloon popper.

The balloon popper will eat batteries, so it is a good idea to use rechargeable cells.

In addition, you will also need the following tools:

TOOLBOX
■ An electric drill and assorted drill bits
■ Epoxy resin glue or a hot glue gun
■ Soldering equipment

Step 1. Drill the Front Panel and Mount the Components

Get all the components together and lay them out next to the box so you can see where everything is

going to fit. Then, mark the lid of the box where you need to drill holes for the variable resistor, phototransistor, terminal block, switch, and LED.

To fix the terminal block, make two small holes, one for each lead, and then glue the block in place (Figure 5-10).

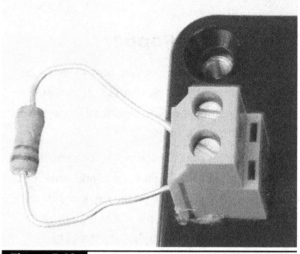

Figure 5-10   Fixing the terminal block in place

PARTS BIN			
Part	**Quantity**	**Description**	**Source**
R1	1	100kΩ variable resistor	Farnell: 1227589
R2	1	1kΩ 0.5-W metal film resistor	Farnell: 9339779
R3	1	2.2Ω 2-W resistor	Farnell: 1565455
R4	1	100Ω 0.5-W metal film resistor	Farnell: 9339760
D1	1	5mm red LED	Farnell: 1712786
T1	1	NPN phototransistor (visible light)	Farnell: 1497673
T2	1	FQP33N10 MOSFET transistor	Farnell: 9845534
Connector	1	Two-way PCB mount terminal block	Farnell: 1386121
Battery clip	1	PP3-style battery clip	Farnell: 1183124
Battery holder	1	4 × AA battery holder with clip terminal	Farnell: 1696782
Batteries	4	AA NiMH rechargeable cells	
Switch	1	DPDT on-off-on toggle switch	Farnell: 9473530
Box	1	Small project box	Farnell: 301243
Balloons			
Knob	1	Plastic knob to fit R1	Farnell: 1441137

Figure 5-11 Components fitted onto the box lid

Fit the switch and variable resistor, tightening their retaining nuts with pliers. Drill a hole into which the LED will snugly fit. If it is a tight fit, it will stay in place, otherwise a drop of glue on the underside of the lid will stop it from moving.

When all the parts are attached, it will look like Figure 5-11. Notice the glue around the leads of the phototransistor to hold it in place.

Step 2. Solder the Other Components

There is no circuit board in the balloon popper. Since there are only a few components, we can simply solder the remaining components to those fixed to the box lid. Use the wiring diagram of Figure 5-12 as a guide.

The short lead of the phototransistor is the collector. This lead should be connected to the switch.

When all the components and wires are in place, you should have something that looks like Figure 5-13.

Step 3. Final Assembly

Inspect everything carefully to ensure no wires are touching. If any wires are very close to each other and may move, wrap insulating tape around them.

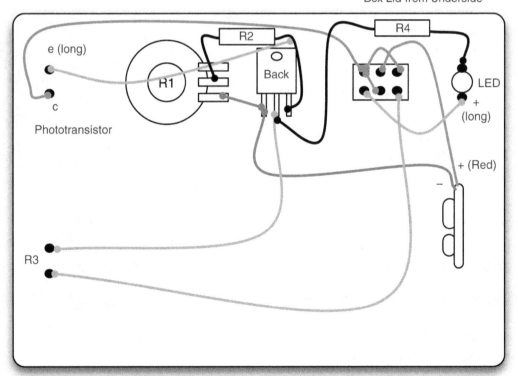

Figure 5-12 The wiring diagram for the balloon popper

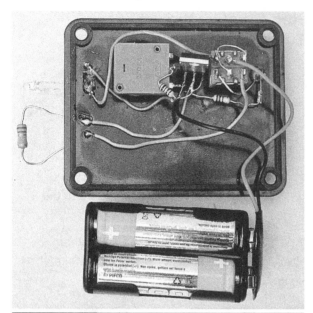

Figure 5-13 The inside of the balloon popper, with all the components in place

Before fitting everything in the box, let's carry out a basic check on the electronics.

Fit the batteries into the battery holder; make sure the switch is in the center "off" position and attach the battery clip. Now move the switch to the "test" position (toward the LED). Next, turn the knob from one end of its travel to the other. You should see the LED turn on at some point. Set the variable resistor so the LED is on, but only just on. Then, move your hand over the phototransistor and the LED should go off.

If this does not work, go back and check your wiring, and make doubly sure that the phototransistor is the right way around.

Once everything is working, fit it all inside the box and screw down the lid (Figure 5-14).

Testing the Balloon Popper

Before we pop our first balloon, we need to check that the resistor gets hot. To do that, set the switch to "test" and turn the knob until the LED comes on. Flip the switch to "live" and hold your finger close to the resistor (don't touch it). After a few seconds, you should feel heat coming from it.

Figure 5-14 Final assembly of the balloon popper

Immediately set the switch back to "off" and wait for it to cool down.

The best way to test the balloon popper is to select the nerviest of your minions and have them stand next to the balloon popper to confirm the balloon has popped.

First attach the balloon to the resistor using Scotch tape, as shown in Figure 5-15.

Figure 5-15 A balloon attached to the popper

Set the switch to "test" (toward the LED) and then adjust the variable resistor until the LED is just off, then turn it a little bit further so it is still off. This is adjusting the sensitivity of the sensor so it is not affected by the ambient light.

Now shine the ray gun at the balloon. When it comes within a certain range of the phototransistor, the LED should light. Practice aiming the laser at a point where the LED will light, since you will need to keep the beam on this point for ten seconds or so to pop the balloon.

When you are confident you can do this, position your minion next to the balloon (as an observer) and then flip the switch to "live." Aim the beam at the balloon. After ten seconds or so, there should be a loud bang and a terrified minion. You should then repeat the experiment from various distances. Resist the temptation to place the balloon popper on the head of the minion, as you may end up lasering the minion's eyes.

Can Shooter

Popping balloons is all well and good, but for added impact the Evil Genius also likes to shoot cans off the wall with his ray gun. Again, there is some deception going on here. The can is a "special" can with some electronics in it. When the beam from the laser hits the phototransistor sensor, a motor swings a weight that hits the side of the can and makes it jump. If the can is carefully balanced, the movement will be enough to make it fall off the surface it is standing on. Figure 5-16 shows the can jumper, which as you might expect just looks like a can full of the Evil Genius' favorite lunchtime snack.

Figure 5-17 shows the schematic diagram for the can jumper.

This uses a light sensor in a similar way to the balloon popper. However, in this case, when the laser hits the phototransistor turning on the MOSFET power transistor, rather than heating a

Figure 5-16 The can jumper

resistor, it turns on a motor that swings an arm around that hits the can.

The diode across the motor prevents the high reverse voltages you get when using a motor from destroying the MOSFET.

What You Will Need

You will need the components for the can shooter that are listed in the Parts Bin.

You do not usually get much information about a small DC motor when buying one. Often, all you get is its nominal voltage. The author found his motor in a local electronics store for a few dollars. You may also find one as part of a scrap toy or an educational motor kit.

Look for a motor about the same size as the one shown in Figure 5-21 later in the chapter. Chances are it should have a similar amount of power. A 4–6V motor should be fine even though we are going to momentarily power it from 9V. It should not have time to burn out for the small fraction of a second it is on. However, do not be tempted to buy a lower voltage than this because you may well burn it out.

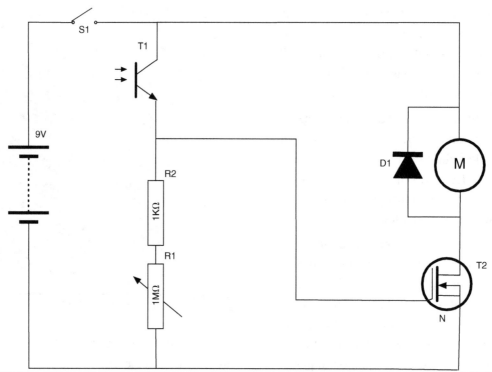

Figure 5-17 The schematic diagram for the can jumper

PARTS BIN			
Part	**Quantity**	**Description**	**Source**
R1	1	1MΩ preset variable resistor	Farnell: 1227571
R2	1	1kΩ 0.5-W metal film resistor	Farnell: 9339779
Motor	1	4-9V DC motor	Farnell: 599128
T1	1	NPN phototransistor (visible light)	Farnell: 1497673
T2	1	FQP7N10 MOSFET transistor	Farnell: 9845534
D1	1	1N4004 diode	Farnell: 9109595
S1	1	Micro-switch	Farnell: 1735350
Perforated board	1	Perforated prototyping board at least 2" × 4" (50 × 100mm)	Farnell: 1172145
Battery clip	1	PP3-style battery clip	Farnell: 1183124
Battery	1	PP3 9V rechargeable NiMH battery	
Can	1	Ring-pull food can (16 oz or 415 g)	
Lid	1	Circle of card diameter to suit the can	
Nail	1	2" × ⅛" diameter (50mm × 2mm)	
Terminal block	1	Five-way 2A screw terminal block	Farnell: 1055837
Diffuser	1	Expanded plastic packaging material	
Screw	1	Self-tapping screw ½-inch (10mm)	

The battery is a rechargeable 9V PP3 battery. There are good reasons for using a rechargeable battery here. The motor is likely to draw several amps, the battery will become dead quite quickly, and it is greener and more economical to recharge rather than replace.

The expanded packing material is going to be used to diffuse the light from the laser. Look for something that will allow light to pass but diffuse it like frosted glass. We also use this useful material in Chapters 7 and 10, which focus on a laser beam alarm and laser voice transmission.

In addition to these components, you will also need the following tools:

TOOLBOX
■ An electric drill and assorted drill bits
■ Epoxy resin glue or a hot glue gun
■ Soldering equipment
■ Scissors
■ Scotch tape

Figure 5-18 shows the overall design behind the can shooter.

All the components, including the motor and battery, are mounted onto a piece of perforated prototyping board. This board is like stripboard, but without the strips of copper. It's just a board with holes drilled into it at a pitch of 0.1 inches. It provides a useful framework to which the components can be attached.

Because the motor swings upward, gravity will take care of returning the arm to its resting position, making it ready for the next shot.

The design has a micro-switch at the top of the board so the can is turned on when the lid is put in place. Since the lid also blocks out the light to prevent the can jumper from being activated by ambient light, this works well.

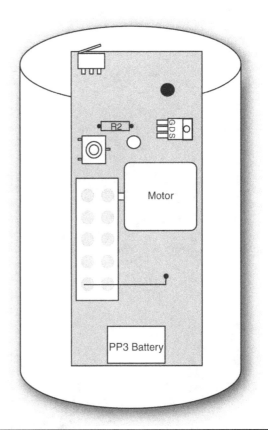

Figure 5-18 Design of the can jumper

Step 1. Cut the Perforated Board

First, find an empty one-pound (450 g) food can. The type with a ring-pull top is needed since it is structurally stronger after it has been opened and has far fewer sharp edges on which to cut yourself. The exact dimensions of cans vary, so you will need to get the length of the perforated board to make an exact fit with your can. The external dimensions of the Evil Genius' can were 4¼" × 3" diameter (108mm with a diameter of 75mm).

Cut the perforated board with scissors so it is the right length to fill the can from top to bottom. The width is less critical, meaning the width of 1¹⁵⁄₁₆ inches (49mm) shown in Figure 5-19 should be fine unless you have a very unusually shaped can. The board should be a snug fit top to bottom.

1 15/16 inch (49mm)

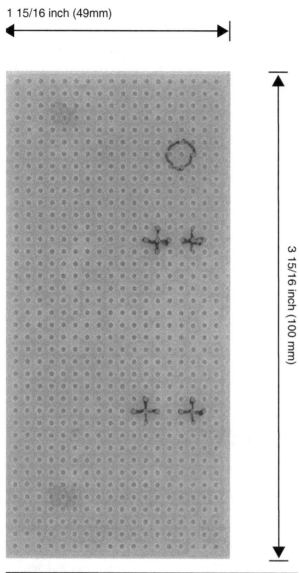

3 15/16 inch (100 mm)

Figure 5-19   Sizing the perforated board

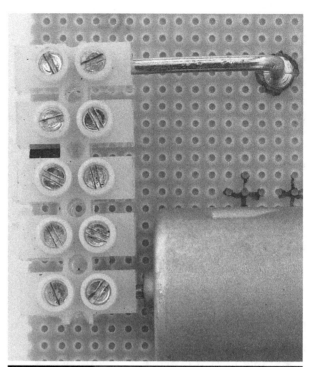

Figure 5-20   The arm assembly

Step 2. Build the Arm Assembly

Electric motors usually just end in a metal shaft. The way we attach the arm to the shaft is to use a screw terminal block as our arm. We can just pop the shaft into one of the terminal connectors and tighten the screw. The terminal block strips usually come with ten or more connections in a strip. We only need five, but they can be easily cut into a section of five, using a knife.

The terminal block is reasonably heavy so as to provide some momentum to be transferred into the

can. We help this along with a nail, which will also strike the screw.

The whole arrangement is shown in Figure 5-20.

One of the nails is going to be bent at right angles about half an inch (12mm) from its end. This will strike the self-tapping screw fitted at the right place on the perforated board. The screw will allow the stop point of the arm to be adjusted so it is past the tipping point and will fall back after the can has been shot.

The other terminals of the block can be used to attach more weight to the arm. There is, however, a trade-off between the amount of weight on the arm and how fast it moves. So experiment to see what suits your motor best.

Step 3. Attach the Motor

Attach the arm to the motor and position the motor halfway up the perforated board. You now need to mark where the bent nail will hit the perforated board, because we are going to make a hole there so we can fit the self-tapping screw.

Be careful when drilling the perforated board, because it is easy to break if the drill "digs in" to one of the perforations. Choose a drill bit that is just slightly smaller than the diameter of the self-tapping screw.

Once the screw is in place, find just the right place for the motor to fit so the nail hits it dead center. Also, put the board into the can, and make sure everything fits and that the arm has enough room for its complete travel, before gluing the motor into place (Figures 5-21 and 5-22).

Before we add the electronics, we need to test the motor to make sure everything we have done so far works. We also need to work out the polarity of the motor so we know which terminal goes to positive and which to negative.

After putting the arm in its resting position, take your 9V battery and very briefly touch the contacts to the motor contacts. If it flies up to the top

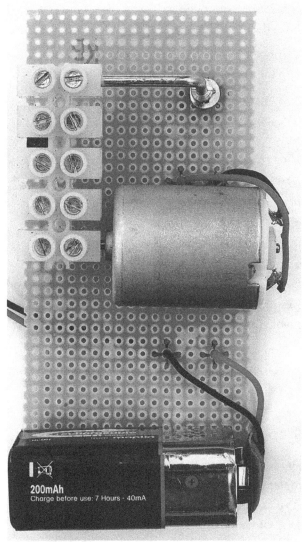

Figure 5-21 The motor attached to the board in the "up" position

Figure 5-22 The motor attached to the board in the "down" position

position, make a note of which terminal on the motor was positive.

If it just kind of twitches a bit, then try the battery the other way around.

At this point, you can also experiment with different weights on the arm to see what works best for your motor.

Step 4. Add the Electronics

We can now fit the components to the perforated board. The component leads are pushed in from the top of the board and connected together on the back of the board using the component leads and wire. Figure 5-23 shows the wiring diagram.

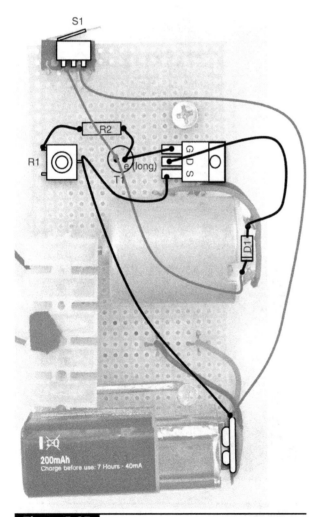

Figure 5-23 Can jumper wiring diagram

Push all of the components into place and then use Figure 5-23 as a reference to connect up the components. The front of the completed board is shown in Figure 5-24, while Figure 5-25 shows the back of the board.

Take particular care to get the phototransistor the correct way around. Check its data sheet, but the longer lead is normally the emitter and the shorter lead the collector. The collector goes to the positive 9V supply.

Glue the battery clip perpendicular to the board, as shown in Figure 5-24.

When connecting wires to the motor, make sure you connect the positive power supply to the connection you found to be positive in the test at the end of the previous section.

We can now test everything before we move onto the final stage of fitting it all into the can.

While in the can, the phototransistor will be in very low light, so we need to do this testing in fairly low light to prevent the phototransistor from being turned on all the time. We can also turn the preset variable resistor to its most clockwise position to make the sensor as insensitive as possible.

Connect everything up and put the arm in its resting place. If the motor is activated, then disconnect the battery immediately and check everything. Remember, if the motor is left on but is unable to move, it may burn out.

Now shine the laser (or any light) onto the phototransistor and the arm should fly up to the top position.

Step 5. Prepare the Can

The first thing we need to do is consume the contents of the can and clean it out thoroughly. If the can contains a food that is not to the liking of the Evil Genius (dog food, for instance), it can be fed to a minion since the Evil Genius believes it's a crime to waste food. The can should be of the

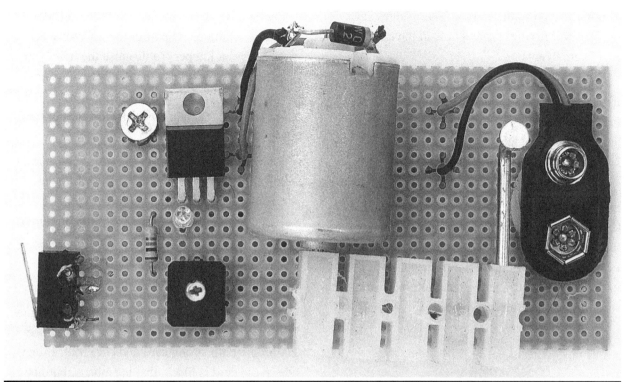

Figure 5-24 The front of the completed board

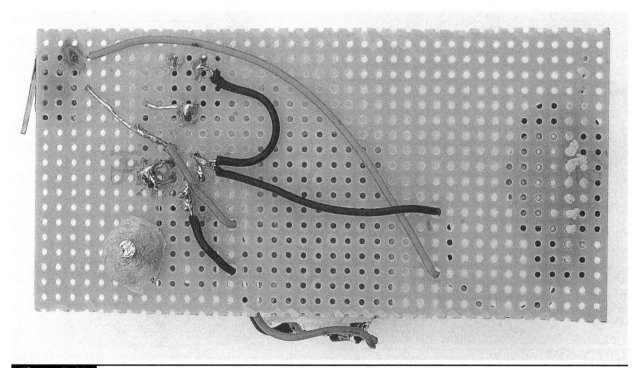

Figure 5-25 The back of the completed board

"ring-pull" type so there are no sharp edges at the top. You will need to make a hole in the can for the laser beam to enter.

Put the board into the can and mark a position directly opposite the phototransistor. Drill a hole there about ¼ inch (6mm) in diameter. Now, tape a pad of the packing material about 1½" × 1" over the hole using Scotch tape (Figure 5-26).

Without the packing material as a diffuser, the laser would pass straight through the hole and only activate the can if a direct hit was scored on the phototransistor. With the packing material in place, the laser illuminates the material, which then lights up the entire can, triggering the phototransistor. You can see this effect in Figure 5-27.

Figure 5-28 shows the hole with the diffuser behind it. A hit on any part of the diffuser will trigger the can jumper. To disguise the hole, you could glue the label back on. A label with some white in the pattern would be ideal so that a hole could be cut in the label over the location of the hole on the can.

It now just remains to fit everything into the can and test the can jumper. The battery sticks out perpendicular to the board and takes up most of the width of the can. This will keep the bottom of the board in place.

The top of the board is held in place with adhesive putty (see Figure 5-29).

We now need to make the lid of the can (Figure 5-30). Using the rim of the can as a guide, trace the outline onto strong card and cut it out. If one side of the card is black, that is a bonus, but not essential. The lid needs to be a snug fit so it stays in place inside the rim of the can and holds the

Figure 5-26 Fitting the diffuser

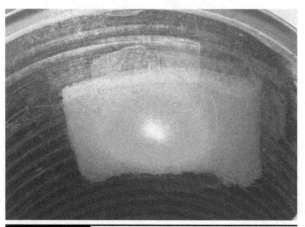

Figure 5-27 The diffuser being illuminated by the laser

Figure 5-28 The hole in the can

Figure 5-29 The can with the board fixed in place

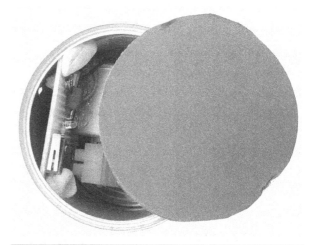

Figure 5-30 The lid to the can jumper

micro-switch in the "on" position. If the card has a black side, this should be on the inside of the lid.

Testing the Can Jumper

With just a small hole to allow ambient light in, the can is going to be quite dark inside, so you should be able to set the sensitivity to maximum. It is best to adjust the sensitivity with the board outside of the can as it is difficult to access the variable resistor when inside the can.

Maximum sensitivity means the variable resistor should be in its most counterclockwise position. So set it to that and confirm everything is okay by pressing the micro-switch. Unless you are working in the dark, the arm will fly up. Now reassemble everything and put the lid on. The can jumper may activate as you put the lid half in place, but you should hear the arm return to its resting place when the lid is fully in place and blocking out the

light. If it does not return without you putting your finger over the sensor hole, you will need to take it apart again and reduce the sensitivity a bit.

You are now ready to use your can jumper. You may find it amusing to place the can near one of your minions and, while out of their sight, activate the can with your laser gun. For best results, select a minion with a fear of the supernatural.

Theory

In this section, we take a look at some of the technology behind this project. In particular, we look at the two types of transistors that we use.

First, we look at MOSFET power transistors used to switch big loads like the the motor, and then we look at the phototransistor that we use to trigger the MOSFET.

MOSFETs

MOSFETs (Metal Oxide Semiconductor Field Effect Transistors) are a type of transistor that is excellent for switching high currents. When they are turned fully off, they have a huge resistance, and when they are on, they have a very low resistance. This means that they generate very little heat when used as a switch that is either fully on or fully off.

They differ from more normal "bipolar" transistors in that they are controlled by voltage rather than current. That is, it is the voltage at the "gate" connection of a MOSFET that determines whether it turns on or not. Only a tiny current flows into the gate, making them ideal for switching from low-current sources like our phototransistor.

Phototransistors

To understand what a phototransistor is, we need to know a little about how a regular bipolar transistor works.

Figure 5-31 illustrates how a regular transistor works.

This transistor has three leads: the emitter, the collector, and the base. The basic principal is that a small current flowing through the base will allow a much bigger current to flow between the collector and the emitter.

Just how much bigger the current is depends on the transistor, but it is typically a factor of 100. So

a current of 10mA flowing through the base could cause up to 1A to flow through the collector and emitter. The actual current flowing will depend on the value of the load resistor. In the case of the balloon popper, this load resistor comprises a variable resistor of 100kΩ and a fixed resistor of 1kΩ. The fixed resistor limits the current when the variable resistor is set to zero.

A phototransistor is a variation on this theme. A phototransistor will often have no base connection and will be contained in a clear plastic case. Instead of a small current being supplied to the base, the current flowing through the collector and emitter is controlled by the intensity of the light falling on the phototransistor.

Summary

It is this kind of trickery that helps the Evil Genius maintain his mastery of the world. Nothing impresses the minions like the threat of instant death by ray gun.

In constructing this project, we have learned a bit about sensing light and controlling power, whether it is to the resistor used as a heating element, or to the electric motor of the can jumper.

In creating this project, an alternative design was considered for the can jumper: using a mechanical mousetrap. These hold quite a lot of energy in them, and they can be triggered by a solenoid. However, in the end, the advantages of a system using a motor that would reset itself was considered the better option. It also seemed more likely that the Evil Genius would retain more fingers by avoiding said mousetraps.

In the next chapter, we will look at adding a touch-controlled laser sight to a BB gun.

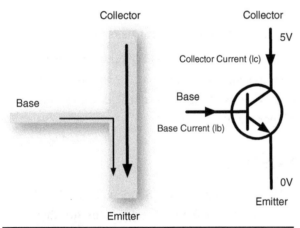

Figure 5-31 The operation of a transistor

CHAPTER 6

Touch-Activated Laser Sight

<table>
<tr><td>PROJECT SIZE:</td><td>Small</td></tr>
<tr><td>SKILL LEVEL:</td><td>★★★☆</td></tr>
</table>

THIS PROJECT ADDS A LASER SIGHT to a BB gun. The laser activates when you touch the trigger prior to firing. This allows the Evil Genius to be far more accurate when carrying out assassinations and other gun-related crimes. Even if it is not connected to the gun, the presence of a bright red dot on the chest of a minion tends to cause an amusing degree of panic.

This project involves a little electronics to make the touch sensor for activating the laser, but this only requires a few components to be connected together. So, if you are looking for a simple project that just requires a bit of soldering to stripboard in order to test your soldering skills, this is an ideal first project for you.

Figure 6-1 shows the completed project attached to a particularly nasty-looking BB gun.

WARNING!

Please take heed of the following safety warning.

- BB guns can be dangerous. Fire them at targets, not at people.

- Never shine a laser into your eyes or anyone else's.

- Resist the temptation to check if the laser is on by peering into it. Always shine it onto a piece of paper or some other light surface.

Figure 6-1 BB-gun laser sight

What You Will Need

To build this project, you will need the components shown in the Parts Bin on the next page.

You can buy laser modules from standard component suppliers like Farnell and RS, but they tend to be very expensive. So look online, where they should cost no more than two or three dollars.

PARTS BIN			
Part	**Quantity**	**Description**	**Source**
D1	1	3-mW red laser LED module	eBay
R1	1	100Ω 0.5-W metal film resistor	Farnell: 9339760
R2	1	10MΩ 0.5-W metal film resistor	Farnell: 1127908
T1	1	2N7000 FET	Farnell: 9845178
Battery clip	1	PP3 battery clip	Farnell: 1650667
Battery	1	PP3 9V battery	Hardware store
Wire for contacts			
Box	1	Small plastic project box	
Terminals	1	Two-way terminal block	Farnell: 1055837
Stripboard	1	Small piece of stripboard; seven tracks, each with four holes	Farnell: 1201473

In addition, you will also need the following tools:

TOOLBOX
■ An electric drill and assorted drill bits
■ Soldering equipment
■ Epoxy resin glue or a hot glue gun

Figure 6-2 shows the schematic diagram for this project.

The laser diode module is controlled by a FET. The gate of the FET is pulled to ground by R2. However, when the two contacts are bridged by the resistance of someone's hand, it increases the voltage at the gate enough to turn on the MOSFET and light the laser.

Assembling the Sight

All the components for this project are built into a small plastic box, with wires that lead to the triggering contacts (Figure 6-3).

Step 1. Drill the Box

Place the battery, terminal block, and laser module into the box in the arrangement shown in Figure 6-4 and mark the outside of the box where you need the light from the laser module to emerge. Drill a ¼-inch (6mm) hole.

We need to make two smaller holes at the other end of the box for the wires of the touch sensor to enter the box and attach to the terminal block. Make the holes just big enough for the two wires. Figures 6-5 and 6-6 show the drilled box.

Step 2. Create the Touch Sensor

The touch sensor uses two contacts, one around the handle of the gun and one on the trigger. The contact on the trigger is made by stripping about an inch and a half of solid core wire and wrapping it around the trigger.

The other contact is wrapped around the handle in a similar way. A bit of tape is used to keep the wire on the handle in place. (See Figure 6-7.)

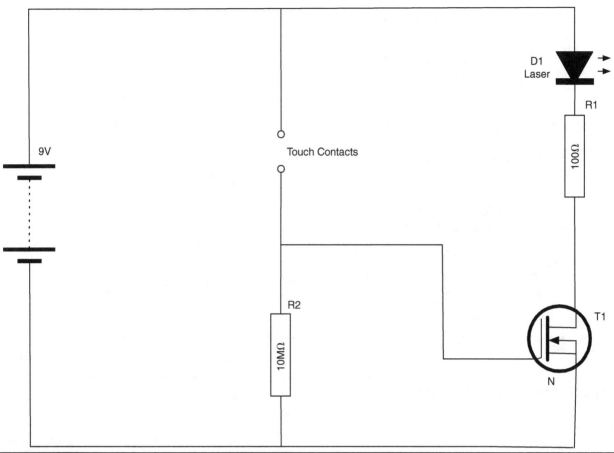

Figure 6-2 The schematic diagram for the BB-gun gunsight

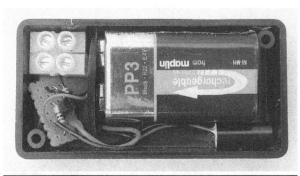

Figure 6-3 The laser sight ready to be attached to the gun

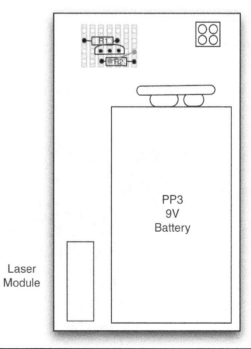

Figure 6-4 The layout of the components in the box

Figure 6-5   The drilled box (front)

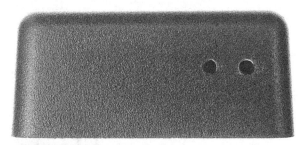

Figure 6-6   The drilled box (rear)

Step 3. Stripboard Assembly

Begin by cutting the stripboard to size. A strong pair of scissors will manage this just fine. Figure 6-8 shows the stripboard layout for the project. It is pretty minimal as there are only three components to put on the board.

First, solder in a short length of solid core wire between tracks 4 and 7.

Solder the two resistors onto the clipboard and then the transistor. Make sure you get the transistor

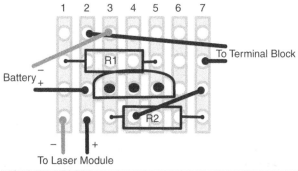

Figure 6-8   The stripboard layout

the right way around. The completed board is shown in Figure 6-9.

Step 4. Fit It into the Box

Lay out the components in the box in the arrangement shown in Figure 6-10, and then shorten the leads of the battery clip and the laser module. They should just reach the point where they will be soldered to the stripboard, with a little bit to spare.

Solder the battery and laser module leads to the stripboard as shown in the wiring diagram of Figure 6-10.

We can now test that it works by connecting the battery and temporarily fitting two wires to the points where the sensor will be attached. When you touch both wires, the laser module should illuminate. It is easier to test everything before you put it in the box.

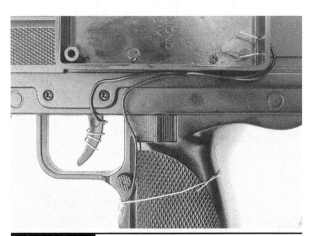

Figure 6-7   The contact wires

Figure 6-9   The completed stripboard

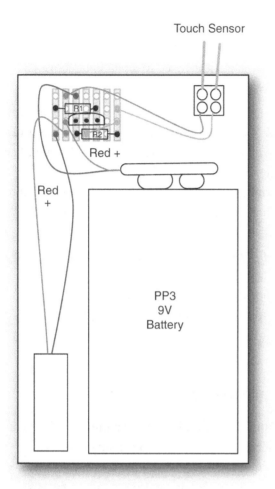

Touch Sensor

Red +

Red +

PP3
9V
Battery

Figure 6-10 The wiring diagram

If everything is all right, we can glue the laser module and terminal strip into place. Use only a small amount of glue so the laser module remains flush with the corner of the box.

Step 5. Finish Up

We can now attach the touch sensor leads to the terminal block and attach the box to the side of the gun. Adhesive putty is a good non-permanent way of attaching the box to the gun. It also allows the angle at which the laser shines to be adjusted a little so it is in line with the gun barrel.

That's it. A nice simple project.

Testing and Calibration

Trial and error is the best way of calibrating the laser. Fix the laser in one position and then fire the gun, noting the position where you hit the target and tweak the laser toward that point.

The laser will never perfectly light up the place where the pellet will hit for a number of reasons, not the least of which is that BB guns do not shoot very straight. And most especially, if shot over a distance, the trajectory of the pellet will be falling, whereas the light from the laser is immune to the effects of gravity.

Theory

The key components of this project are the laser diode and the FET transistor. We now take a detailed look at these two componets.

Laser Diodes

The word laser was originally written as LASER because it stands for Light Amplification by Stimulated Emission of Radiation, but these days it is just written as "laser."

Not that long ago, lasers used to be something you only found in laboratories. They create a beam of light at exactly one frequency (color). They also have the property of coherence, which means that all the waves of the light are in step and don't overlap each other. This makes it possible to use a lens to focus them into a very narrow beam that remains a small dot even when aimed at something quite distant.

Lasers have become cheap and easy to obtain because of the laser diode, which is a similar technology to regular LEDs but generates laser light rather than ordinary incoherent light. These are used in CD and DVD players.

A laser diode on its own is not enough to produce the narrow beam of light we want. We need a lens to do that. Since it is quite difficult to get the lens lined up in exactly the right place, the solution is to buy a laser module. A laser module consists of both a laser diode and lens in one sealed unit with two leads coming out the back to supply it with power. Like laser modules that shine a dot, certain varieties shine a cross or a line. You can also now buy green laser modules. Blue laser modules are also becoming available as a result of Blu-ray disk players, which use high-powered blue lasers.

The laser module we are using is very low power—only 3 mW. However, it is possible to buy laser diodes of 300 mW or more, which when focused are capable of popping a balloon and even burning and cutting. Clearly, such devices must be handled with care and respect.

FETs

You may have noticed that our design does not have an on/off switch. This is unnecessary because the transistor we use is a type called a FET (field effect transistor). These transistors differ from the more common bipolar transistors because, rather than having a "base" connection, they have a so-called "gate." The gate is insulated from the rest of the transistor through which the main current flows. It exerts its influence over the current through the transistor by way of the charge on it. In other words, unlike an ordinary bipolar transistor, which relies on a current to switch the transistor, the FET can be switched just by the presence of potential at the gate. To turn on the current and have the laser light up, simply increase the voltage at the gate to an amount above its threshold voltage.

For the 2N7000, this threshold voltage is around 2V. When the transistor is turned on, it has a very low resistance (typically 1.2Ω). On the other hand, when it is switched off there is a tiny "leakage" current of around $1\mu A$, or a millionth of an amp, which is the reason we don't need an on/off switch.

In Chapter 5, we encountered a type of FET called the MOSFET, which was employed to control the much higher power of a motor and resistor used as a heating element in the balloon popper.

Summary

In this simple project, we explored how to use a laser diode and FET. You can adapt the basic circuit to make other things "touch-controlled" as well. The FET can control a current of up to 400mA. This is more than enough to power a high-brightness LED (HB LED). So you may feel the urge to experiment and see what else you can do with this circuit.

In the next chapter, we will move away from weapons, toward security, starting with another laser project. The laser beam alarm will use mirrors and another laser module to protect the Evil Genius' Lair.

CHAPTER 7

Laser-Grid Intruder Alarm

PROJECT SIZE:	Medium
SKILL LEVEL:	★★★☆

LASER BEAMS LINE THE ENTRYWAY to the Evil Genius' Lair (Figure 7-1), forcing the secret agent to bend and arch through the beams in a series of agile moves. With skills honed by years of Twister and limbo dancing, the agent navigates his way through the beams. Suddenly an alarm sounds and vertical doors slam to the ground, trapping the agent. The Evil Genius triumphs again.

This is a working alarm. It uses a laser beam that bounces between mirrors attached to two walls that face each other. For instance, the entrance to the Evil Genius' Lair. When the beam is broken, a buzzer sounds.

Note that the more ingenious agent will work out that they can defeat the alarm by attaching a flashlight to the sensor, or indeed just turning it off. Figure 7-1 has also been Photoshopped to show the path of the beam.

The Evil Genius also likes to amuse himself by holding "Laser Limbo" competitions at his sadly under-attended parties. To shouts of "How low can you go?" the Evil Genius limbos under the beam. The wise minion will hide when the buzzer sounds.

Figure 7-1 Laser beam alarm

WARNING!

You should always take certain precautions when using lasers:

- Never shine a laser into your eyes or anyone else's.

- Resist the temptation to check if a laser is on by peering into it. Always shine it onto a sheet of paper or some other light surface.

What You Will Need

You will need the following components to build this project.

You can buy laser modules from standard component suppliers like Farnell and RS, but they tend to be very expensive. So look online, where they should cost no more than two or three dollars.

The mirrors were bought in a pack of six from a hobby and craft retailer.

The plastic packing material should allow light to pass through it, but it will diffuse the light like frosted glass.

In addition, you will also need the following tools, shown in the Toolbox.

Figure 7-2 shows the schematic diagram for this project.

PARTS BIN			
Part	**Quantity**	**Description**	**Source**
Laser	1	5 mW red laser LED module	eBay
R1	1	1MΩ variable resistor	Farnell: 1629581
R2	1	1MΩ 0.5-W metal film resistor	Farnell: 9339809
R3	1	270Ω 0.5-W metal film resistor	Farnell: 9340300
R4	1	100Ω 0.5-W metal film resistor	Farnell: 9339760
T1	1	NPN phototransistor (visible light)	Farnell: 1497673
D1	1	5mm red LED	Farnell: 1712786
T2	1	2N7000	Farnell: 9845178
Power socket	1	2.1mm power socket	Farnell: 1217038
Power supply	1	12V 0.5A (or more) power supply with 2.1mm connector	Farnell: 1279478
S1	1	DPDT on-off-on toggle switch	Farnell: 9473530
Box	1	Small plastic project box	Farnell: 301243
Buzzer	1	Piezo buzzer	Farnell: 1192513
Knob	1		Farnell: 1441137
Mirrors	3	Small mirrors	Craft shop
		Adhesive putty	
Diffuser		Expanded-plastic packing material, about ½-inch square (10mm)	Recycling
Cable	6 feet	Twin core flex (1800mm)	Hardware store

TOOLBOX

- An electric drill and assorted drill bits
- Soldering equipment
- Glue

The circuit is very similar to the balloon popper part of Chapter 5. However, in this case we want to know when the beam is interrupted. For this reason, the phototransistor is connected as a "common emitter"—that is, the emitter of the transistor is connected to ground rather than the collector being connected to the positive supply as in the balloon popper.

Assembly

The following step-by-step instructions lead you through making the laser alarm. It is quite a simple design, so we can actually build it without a circuit board.

Step 1. Drill the Control Box Lid

Most of the holes for this project are going to be made in the lid of the project box. For the sake of neatness, it's a good idea to line them all up along the center of the box. So start using a ruler and pencil to draw a line on the inside of the lid from one side to the other across its long side.

Next, lay out the LED, switch, variable resistor, and the expanded plastic packing material that you are going to use as a diffuser. Make sure there is enough room around each of these components and mark the point on the line where you need to drill. Drill holes for all the components. Drill a relatively large hole that is slightly smaller than the diffuser. The author also enlarged and squared off this hole with a file. However, this is not essential.

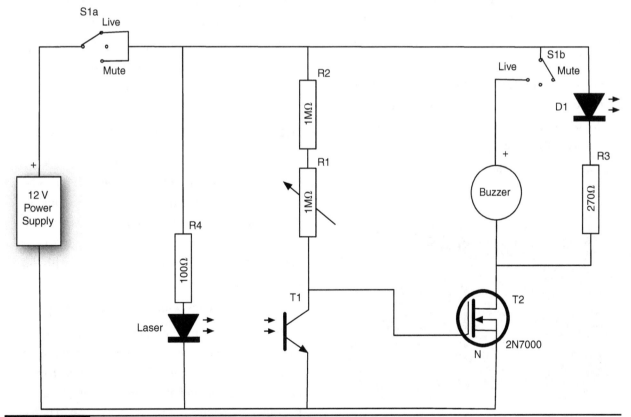

Figure 7-2 The schematic diagram for the laser alarm

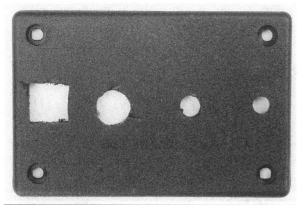

Figure 7-3 The box lid, drilled

Figure 7-3 shows the lid drilled and ready to accept the components.

You can now attach the LED, switch, and variable resistor. If you drill the hole for the LED with a 5mm drill bit, the 5mm LED will fit tightly and stay in place without the need for a holder.

Step 2. Drill the Control Box

We must now drill in the control box itself. We need a hole to accept the 2.1mm power socket, two holes for the connections to the buzzer, and a hole for the lead to the laser.

Plan out the placement of the holes so they are well away from the other components on the lid. Before drilling the holes, put the lid on to make sure everything fits in the box.

Figure 7-4 shows the box itself drilled.

Figure 7-4 The project box, drilled

Figure 7-5 The project box with buzzer and socket fitted

Push the leads of the buzzer through the two holes and glue the buzzer into place on the side of the box (Figure 7-5).

Step 3. Wire Up the Lid

Using Figure 7-6 for guidance, wire up the components on the lid.

Figure 7-7 shows the box lid with the main components attached, while Figure 7-8 shows some of the other components attached.

Bend the wires on the phototransistor so it faces the diffuser. Attach leads that will connect to the power connector and buzzer in the box.

Step 4. Wire Up the Laser Module

Solder the leads from the laser module to one end of the six-foot (1.8m) twin core cable (Figure 7-9). If possible, use cable that has a stripe or some way of indicating which core is to be used as positive. If the cable is unmarked, use a multimeter to ensure that the laser module is wired up with the correct polarity.

Insulate the leads with PVC insulating tape.

To relieve the strain on the cable, we are going to thread the cable through the hole in the project box and then tie a knot in it. We can then strip the ends of the cable and attach the negative lead to the outer connection of the power socket. The

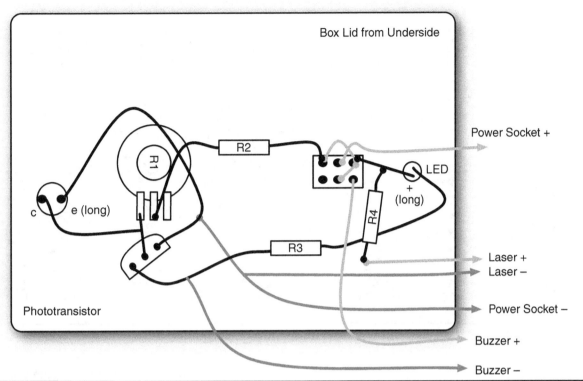

Figure 7-6 Wiring diagram

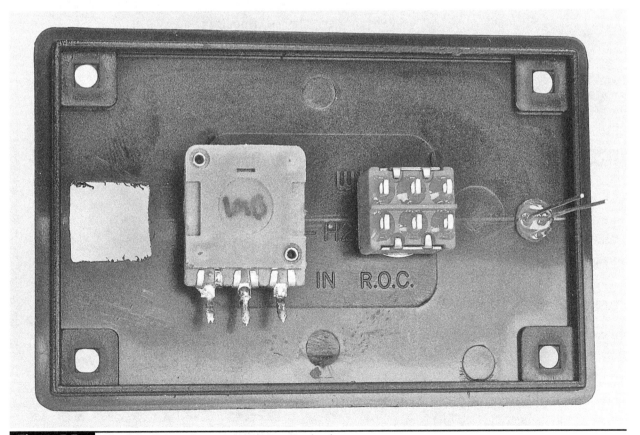

Figure 7-7 The box lid, main components attached

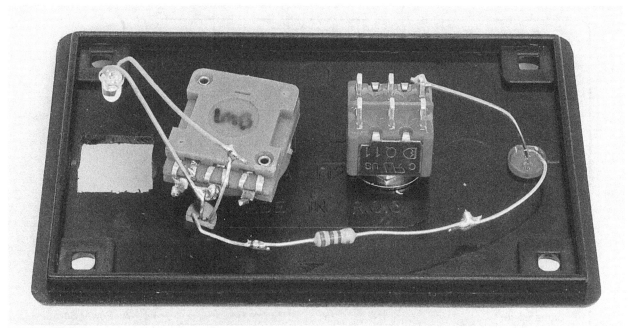

Figure 7-8 The box lid, partially assembled

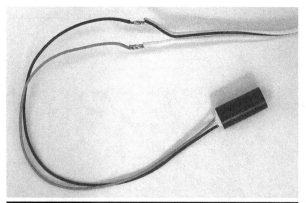

Figure 7-9 Attaching the laser to the cable

positive connection will attach to the 100Ω resistor R4, whose leads should be cut short. Attach a short length of multi-core wire to the other end of the resistor ready to be connected to the switch.

Step 5. Complete the Wiring

We can now wire up the box itself and connect it to the lid.

Start by attaching the positive and negative power leads to the appropriate leads of the power socket. Always solder the lead that is most inaccessible first, because it will only become

less easy to reach as other things are soldered into place.

Next, attach the leads to the buzzer and finally the lead from R4 to the switch.

Figure 7-10 shows the completed wiring.

Figure 7-10 The completed wiring

Testing

Before we screw down the box lid, we should try a simple test to make sure that everything is working. You will probably want to do this with the switch in the "mute" position.

Attach the power supply and turn it on. Cover up the diffuser so it is dark. This should cause the LED to light. If the LED was already lit when the power was turned on, try shining the laser module at the diffuser. The LED should go out. If this does not happen, then something is wrong and you need to go back and check your circuit.

You can now fasten the box together and adjust the variable resistor until the LED is just on. Shine the laser onto the diffuser and it should go out. Then, move the beam away from the diffuser and the LED should light as if the beam had been broken.

The completed control box is shown in Figure 7-11.

Installation

We now come to the installation of the alarm. This requires us to fix the laser module to the wall, as well as the three mirrors that are going to direct the path of the laser. Figure 7-1 will remind you of the way the alarm is laid out.

To do this, we are going to have the laser turned on and then place the mirrors rather than try to fix the mirrors in place first.

Start by fitting a large lump of the adhesive putty around the laser module and fit it at about head height to the wall, pointing down at an angle of about 30 degrees (Figure 7-12).

Please note that while the Evil Genius is not too concerned about the décor of his lair, you might be, and the use of adhesive putty on painted walls generally causes a mess and the loss of some of the paint when you finally remove the putty.

Using a small bit of the putty, attach one of the mirrors on the wall opposite the laser about one third of the way to the ground. This mirror should

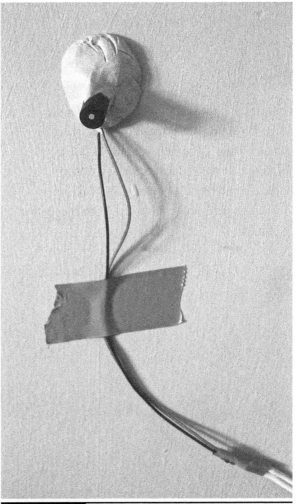

Figure 7-12 Fixing the laser module to the wall

Figure 7-11 The completed control box

be flush to the wall. Adjust the laser, moving it in the putty until it hits as near the center of the mirror as possible.

Now fix the next mirror where the dot appears on the wall opposite. Repeat this for the last mirror, which should be attached to the wall with a bigger bit of putty to allow it to be angled slightly back to the control box.

The control box should be fitted so the laser hits the diffuser. At this point, the LED should turn off and the alarm is successfully installed.

This is a project that requires a little trial and error to get right. You may find that your laser is bright enough and your mirrors good enough to achieve more bounces over a greater distance.

You may prefer not to use a mirror at all but instead protect a longer region with a single beam pointing directly into the control box.

Theory

This alarm is interesting in that it shows the difficulty of accurately positioning a laser and mirrors. Walls are never completely flat, nor absolutely vertical. Mirrors are not perfect reflectors and the laser beam itself will not travel in a tight beam forever before it loses much of its brightness.

Although when you shine a laser on a distant wall at night it may seem like lasers travel for a tight beam forever, actually they don't. The type of coherent light they use is easy to focus, but it must actually be focused and the laser module includes a glass or plastic lens to accomplish this.

Lenses are never perfect and hence the beam, although very narrow, will eventually diverge. The more it diverges, the more spread out the light and therefore the dimmer it is where it hits a small sensor like our phototransistor.

Summary

An improvement on the design would be to use a bistable latch, so that once triggered, the buzzer does not stop sounding until the alarm is reset. If you search the Internet for "bistable latch," you should find some resources and an explanation for the adventurous reader on how to try this modification.

In the next chapter, we are going to meet an Arduino microcontroller board and use it to create an impressive spinning persistence-of-vision effect.

CHAPTER 8

Persistence-of-Vision Display

PROJECT SIZE:	Large
SKILL LEVEL:	★★★★

ATOP THE EVIL GENIUS' LAIR, there is a sign that is holographic in appearance and displays the slowly rotating message "Hello." The sign uses a vertical line of multicolor LEDs that rotates rapidly, producing a persistence-of-vision effect (see Figure 8-1). The main purpose of such a display is to help the Evil Genius find his way back to the lair, exhausted after a hard day achieving world domination.

Figure 8-1 Persistence-of-vision display

This project has two parts, an Arduino microcontroller-based LED unit that sits on top of the second part of the project: the motor and motor control unit. This controls the speed of the motor that spins the arm with the LEDs.

WARNING!

This project has two potentially hazardous aspects.

- **It spins around quite fast, so if you stick your fingers into the path of the wooden arm while it is spinning, it will hurt!**
- **The entire project is based on bright flashing lights. If you suffer from epilepsy, then steer clear of it.**

Arduino

Central to this project is the Arduino microcontroller board (Figure 8-2). We will use these boards again in Chapters 13 and 15.

Arduino interface boards provide the Evil Genius with a low-cost, easy-to-use technology to create their evil projects.

Arduino is a small microcontroller board with a USB plug to connect to your computer and a number of connection sockets that can be wired up to external electronics such as motors, relays, light sensors, laser diodes, loudspeakers, microphones, and so on. They can either be powered through the USB connection from the computer, or from a 9V battery, and can be controlled from the computer

Figure 8-2 An Arduino board

or programmed by the computer and then disconnected and allowed to work independently.

At this point, the Evil Genius might be wondering which top secret government organization they need to break into in order to acquire an Arduino.

Alas, no evil deeds are necessary to acquire one of these devices. The Evil Genius need go no further than their favorite online auction site or search engine. Since the Arduino is an open-source hardware design, anyone is free to take the designs and create his own clones of the Arduino and sell them—thus, the market for the boards is competitive. An official Arduino costs about US$30 and a clone often less than US$20.

The name Arduino is reserved by the original makers of the Arduino. Clone Arduino designs often have the letters "duino" on the end of their name—for example, Freeduino or DFRduino.

The software for programming your Arduino is easy to use and also freely available for Windows, Mac, and LINUX computers, at no cost.

Although Arduino is an open-source design for a microcontroller interface board, it is actually more than that since it encompasses the software development tools you need to program an Arduino board, as well as the board itself. There is a large community of construction, programming, electronics, and even art enthusiasts willing to share their expertise and experience on the Internet.

In this book, we use the Arduino Duemilanove, sometimes called Arduino 2009; however, the recently released Arduino Uno will also work just fine.

When making a project with an Arduino, you need to download programs onto the board using a USB lead between your computer and the Arduino.

This is one of the most convenient things about using an Arduino. Many microcontroller boards use separate programming hardware to get programs into the microcontroller. With Arduino, it's all contained on the board itself.

On both edges of the board are two rows of connectors. The row at the top of the diagram is mostly digital (on/off) pins, and any marked with "pwm" can be used as analog outputs.

The bottom row of connectors has useful power connections on the left, and analog inputs on the right.

These connectors are arranged like this so that so-called "shield" boards can be plugged into the main board in a piggyback fashion. It is possible to buy readymade shields for many different purposes. The following are just a few of the things they are used for.

- Connection to Ethernet networks
- LCD displays and touch screens
- XBee (wireless data communications)

- Sound
- Motor control
- GPS tracking

Searching the Internet for "Arduino Shields" should produce lots of interesting results if you want to go further into the world of Arduino. In fact, one book in the Evil Genius series by this author concentrates solely on the Arduino. Its title: *30 Arduino Projects for the Evil Genius*.

Persistence-of-Vision Display

We are going to start by assembling the LED and Arduino module. The finished module is shown in Figure 8-3 and its schematic diagram in Figure 8-4.

The LED module uses nearly all the input/output pins available on the Arduino board in order to drive each of the LEDs separately. Each LED has a current limiting resistor, which is 150Ω for the red LEDs and 100Ω for the other colors.

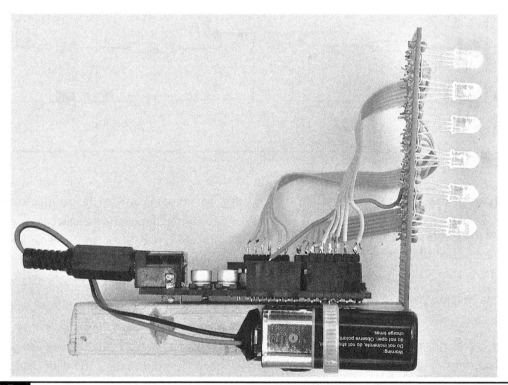

Figure 8-3   The LED module

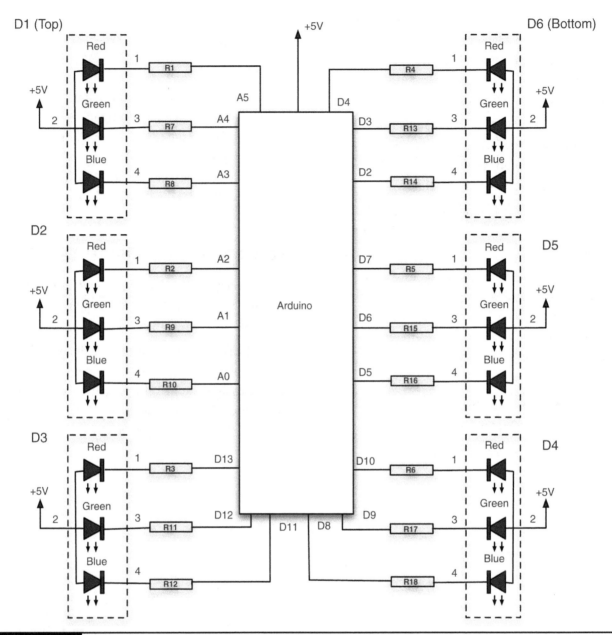

Figure 8-4 Schematic diagram for the LED module

What You Will Need

You will need the following components to build the Arduino-controlled LED module. They are listed in the Parts Bin.

Later in this chapter, you will find the component listing for the motor controller.

The LEDs can be quite expensive when bought from a regular component supplier. You can get a better deal on eBay. When buying them, make sure

they are common anode (the positive connections of each of the three LEDs connected together) and that the pin connections are the same as indicated in the schematic diagram of Figure 8-4. It is always worth checking the LED's data sheet.

LEDs with a diffuse lens produce a better effect than those with a clear lens. However, if only clear-lens LEDs are available, you can gently roughen the front of the LED with sandpaper to give it a frosted effect.

PARTS BIN			
Part	**Quantity**	**Description**	**Source**
	1	Arduino Uno or Duemilanove	Internet, Farnell: 1813412
D1-6	6	RGB common anode LED—diffuse lens matching the pinout shown in Figure 8-5	eBay
R1-6	6	150Ω 0.5-W metal film resistor	Farnell: 9338489
R7-18	12	100Ω 0.5-W metal film resistor	Farnell: 9339795
Ribbon cable	1	IDE hard-disk cable	Farnell: 778710
Battery clip	1	PP3 battery clip	Farnell: 1650667
Power plug	1	2.1mm power plug	Farnell: 1200147
Stripboard		39 strips each of 12 holes	Farnell: 1201473
Headers	1	18-header pin strip split into three six-way headers	Farnell: 1097954
Battery	1	PP3 battery (rechargeable)	
Wood	1	Piece of wood 4" (100mm) × 1½" (35mm) × ⅝" (15mm)	Hardware store
Cable tie	1	6 inches (150mm)	Hardware store

To connect the LEDs to the Arduino board, we will use ribbon cable. A great source of ribbon cable is an IDE-style hard-disk cable scavenged from an old computer. Even if you have to buy a new cable, the sheer volume of production of these cables means they are much cheaper than buying ribbon cable by itself.

You will also need the following tools:

TOOLBOX
■ Soldering equipment
■ An electric drill and assorted drill bits
■ Wood saw
■ A computer to program the Arduino
■ A USB-type A-to-B lead (as used for printers)

Step 1. Prepare the Stripboard

Figure 8-5 shows the stripboard layout for the LED module.

The first step is to cut a piece of stripboard that has 39 strips each of 12 holes. You must then make 18 breaks in the tracks. Do this by using a drill bit—that is, employing it as a hand tool—rotating it between finger and thumb. Figure 8-6 shows the prepared board, ready to be soldered with the components.

You will also need to drill two larger holes at the bottom of the stripboard to attach it to the wooden arm.

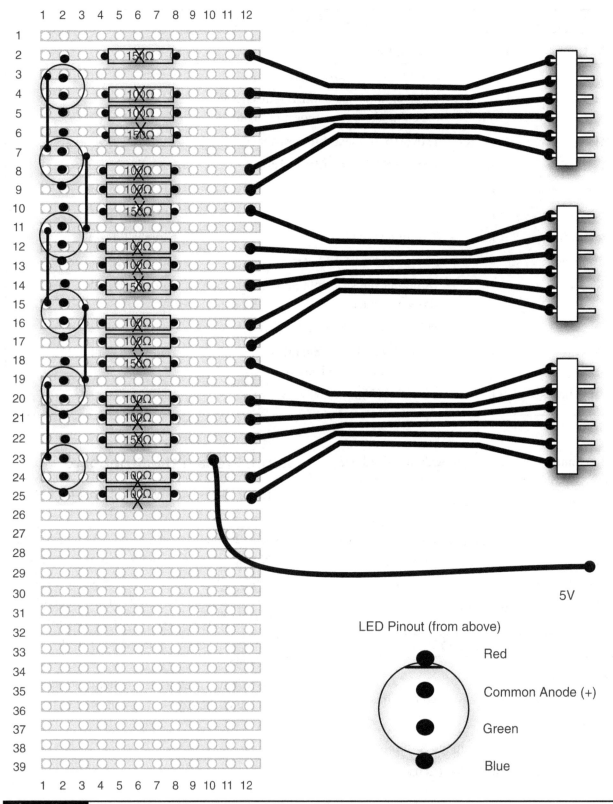

Figure 8-5 The stripboard layout for the LED module

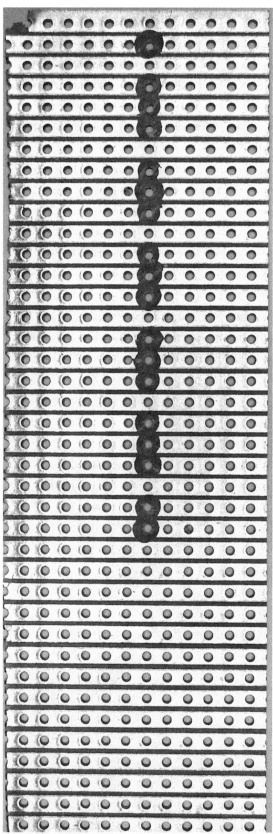

Figure 8-6   The LED module stripboard ready for soldering

Step 2. Solder the Links and Resistors

Before starting on the real components, solder the five wire links into place.

Afterward, solder all the resistors in place. Once done, your board will look like Figure 8-7.

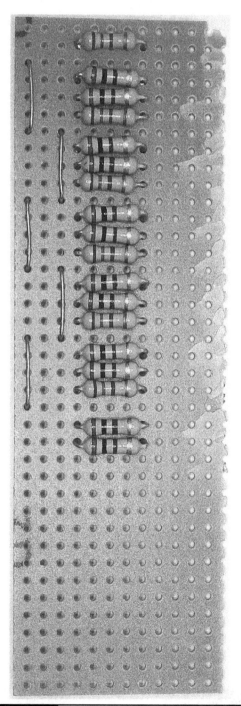

Figure 8-7   The board with resistors and links

Step 3. Solder the LEDs

Place the LEDs into position, ensuring they are the correct way around. Then, starting at one end of the board, solder each LED into place. To keep the LEDs as straight as possible, solder one lead of the LED, then reposition the LED from the front if required and solder the other leads into place.

When all the LEDs are soldered into place, the board should look like Figure 8-8.

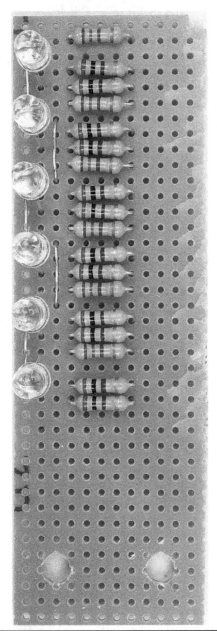

Figure 8-8   The board with all components in place

Step 4. Solder the Ribbon Cable

We are going to use three lengths of ribbon cable, each with six wires. At the stripboard end, the cables will be soldered into place, and at the other end, the ribbon cable will be soldered to header pins so we can easily connect and disconnect the Arduino board.

Begin by cutting the ribbon cable into three lengths each of six wires; the lengths should be roughly 4½ inches (120mm), 3½ inches (90mm), and 2 inches (50mm). Then, strip the ends of all the wires. This is a tedious job as there are 36 ends to be stripped. When all the ends are stripped, they need tinning with solder.

Solder each lead to a header strip (Figure 8-9) and then push the wires on the other end of the board through from the front of the stripboard and

Figure 8-9   Connecting the header strip

solder the lead in place, using the diagram from Figure 8-5 as a guide.

Once all the cables are attached, it only remains to solder a wire for the +5V supply to the stripboard. We used solid core wire attached to the back of the stripboard for this. The other end of the wire can then be pushed into the Arduino board without the need for a header pin.

Step 5. Connect Up

We can now try out our hardware. Connect the pin headers and +5V lead to the Arduino board using Figure 8-9 as a guide.

While testing, we can power the unit from the USB connection to the computer, so attach the Arduino to your computer.

Step 6. Set Up Your Computer with Arduino

To be able to program our Arduino board with the control software for the LED module, we first need to install the Arduino development environment on our computer.

The exact procedure for installing the Arduino software depends on what operating system you use on your computer. But the basic principal is the same for all.

1. Install the USB driver, which allows the computer to talk to the Arduino's USB port. It uses this for programming and sending messages.

2. Install the Arduino development environment, which is the program you run on your computer that enables you to write programs and download them to the Arduino board.

The Arduino web site (www.arduino.cc) contains the latest version of the software.

Installation on Windows

Follow the download link on the Arduino home page (www.arduino.cc) and select the download for Windows. This will start the download of the zip archive containing the Arduino software, as shown in Figure 8-10. You may well be downloading a more recent version of the software than the version 21 shown. This shouldn't matter, but if you experience any problems, refer back to the instructions on the Arduino home page where you will find the most up-to-date information.

The Arduino software does not distinguish between different versions of Windows. The download should work for all versions, from Windows XP onwards. The following instructions are for Windows XP.

Select the Save option from the dialog and save the zip file onto your desktop. The folder contained in the zip file will become your main Arduino directory, so now unzip it into C:\Program Files\Arduino.

You can do this in Windows XP by right-clicking the zip file to show the menu of Figure 8-11 and selecting the Extract All... option. This will open the Extraction Wizard, shown in Figure 8-12.

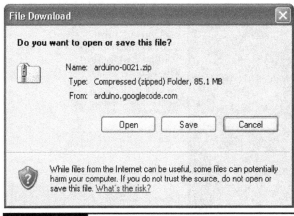

Figure 8-10 Downloading the Arduino software for Windows

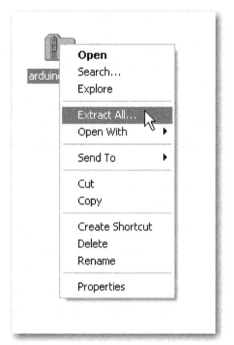

Figure 8-11 The Extract All menu option in Windows

Click Next and then modify the folder to extract files to be C:\Program Files\Arduino, as shown in Figure 8-13. Click Next again.

This will create a new directory for this version of Arduino (in this case, 21) in the folder C:\Program Files\Arduino. This allows you to have multiple versions of Arduino installed at the same

Figure 8-12 Extracting the Arduino file in Windows

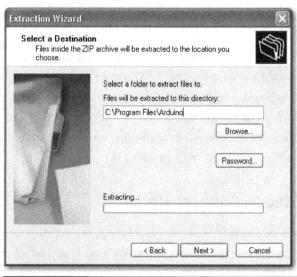

Figure 8-13 Setting the directory for extraction

time, each in its own folder. Updates of Arduino are fairly infrequent and historically have always kept compatibility with earlier versions of the software. So, unless there is a new feature of the software you want to use, or you have been having problems, it is by no means essential to keep up with the latest version.

Now that we have got the Arduino folder in the right place, we need to install the USB drivers.

If you are using the Arduino Uno, this process is a bit different from the older Duemilanove and Diecimila boards.

If you have an Uno board, plug in your board and wait for Plug and Play to fail. Afterward, open the Control Panel from the Start menu. Go to System and then Device Manager, and under COM and LPT ports, you will find the Arduino Uno. Right-click, selecting Update Driver, and navigate to the driver, which you will find in C:\Program Files\Arduino\arduino-0021\drivers\.

For the other boards, we can let Windows help us by plugging in the Arduino board to trigger the Windows Found New Hardware Wizard, shown in Figure 8-14.

Figure 8-14 Windows Found New Hardware Wizard

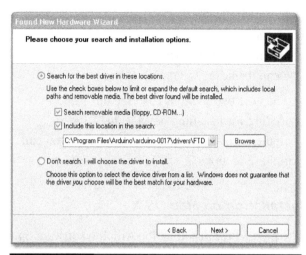

Figure 8-15 Setting the location of the USB drivers

Select the option "No, not this time," and then click Next.

On the next screen (Figure 8-15), click the option to install from a specified location, enter or browse to the location C:\ProgramFiles\Arduino\ arduino-0021\drivers\FTDI USB Drivers, and then click Next. If you download a newer version, you

will have to change "0021", in the path above, to the version you downloaded.

The installation will then complete and you are ready to start up the Arduino software itself. To do this, go to My Computer, navigate to C:\Program Files\Arduino\arduino-0021, and click the Arduino icon, as shown in Figure 8-16. The Arduino software will now start.

Figure 8-16 Starting the Arduino software from Windows

Note that no shortcut is created for the Arduino program, so you may wish to select the Arduino program icon, right-click it, and create a shortcut you can then drag to your desktop.

The next two sections describe this same procedure for installing on Mac and LINUX computers, so if you are a Windows user, you can skip these sections.

Installation on Mac OS X

The process for installing the Arduino software on the Mac is a lot easier than on the PC.

As before, the first step is to download the file. In the case of the Mac, it is a disk image file. Once downloaded, it will mount the disk image and open a Finder window, as shown in Figure 8-17. The Arduino application itself is installed in the usual Mac way by dragging it from the disk image to your Applications folder.

The disk image also contains an installer package for the USB drivers. If you are using an Arduino Uno, you do not need this. For older boards, just perform the following instructions.

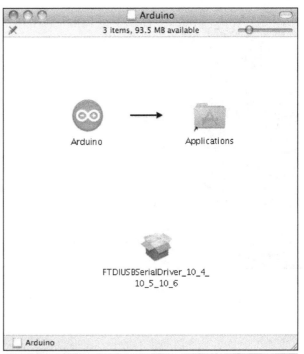

Figure 8-17   Installing the Arduino software on Mac OS X

When you run the installer, simply click Continue until you come to the select disk screen where you must choose the hard disk before clicking Continue again. (See Figure 8-18.)

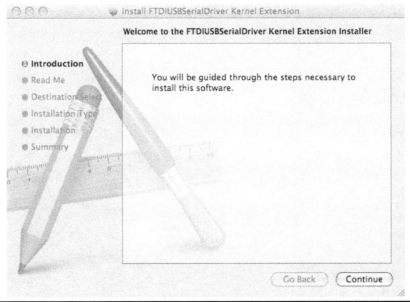

Figure 8-18   Installing the USB drivers on Mac OS X

You can now find and launch the Arduino software in your Applications folder. Since you will use it frequently, you may wish to right-click its icon in the dock and set it to Keep In Dock.

You can now skip the next subsection, which is for installation on LINUX.

Installation on LINUX

Many different LINUX distributions exist, and for the latest information, refer to the Arduino home page. However, for most versions of LINUX, installation is very straightforward. Your LINUX will probably already have installed the USB drivers, the avr-gcc libraries, and the Java environment that the Arduino software needs.

So, if you are lucky, all you will need to do is download the tgz file for the Arduino software from the Arduino home page (www.arduino.cc), extract it, and this will be your working Arduino directory.

If, on the other hand, you are unlucky, then as a LINUX user you are probably already adept at finding support from the LINUX community for setting up your system. The prerequisites you will need to install are Java Runtime 5 or later and the latest avr-gcc libraries.

Googling the phrase "Installing Arduino on SUSE LINUX," or whatever your distribution of LINUX is, will no doubt find you lots of helpful material.

Configuring Your Arduino Environment

Whatever type of computer you use, you should now have the Arduino software installed on it. You must next make a few settings. You need to specify the port that is connected to the USB port for communicating with the Arduino board and we must specify the type of Arduino board we are using. But first you need to connect your Arduino to your computer using the USB port or you will not be able to select the serial port.

The serial port is set from the Tools menu, as shown in Figure 8-19 for the Mac and Figure 8-20 for Windows—the list of ports for LINUX is similar to the Mac.

If you use many USB or Bluetooth devices with your Mac, you are likely to have quite a few options in this list. Select the item in the list that begins with dev/tty.usbserial.

On Windows, the serial port can just be set to COM3.

We can now select the board we will use from the Tools menu, as shown in Figure 8-21. Select the option for the board you are using.

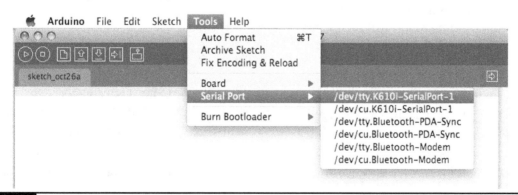

Figure 8-19   Setting the serial port on the Mac

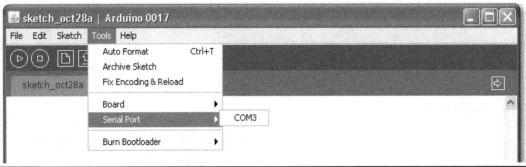

Figure 8-20 Setting the serial port on Windows

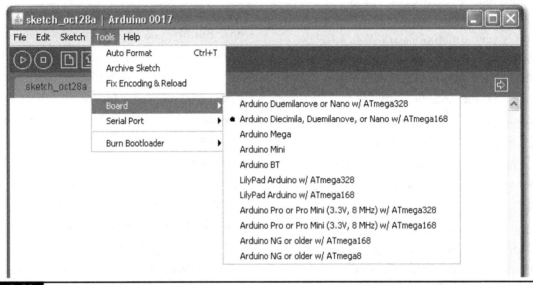

Figure 8-21 Setting the board

Step 7. Program the Arduino Board

Without connecting the power supply to the board, connect a USB cable between your computer and the Arduino board. You should see the red power LED come on, and if it is a new board, a small LED in the middle of the board will be flashing slowly.

Two Arduino sketches for this project are provided at www.dangerouslymad.com: povtest and pov. Install povtest by launching the Arduino software and pasting the code for povtest, which can be found at the web site, into a new project window (see Figure 8-22). Save the project to a convenient location on your computer.

To actually upload the software onto the Arduino board so it can function without the computer, click the Upload button on the toolbar. There will now be a furious flashing of little lights on the board as the program is installed on the Arduino board.

Step 8. Test the LED Module

When the Arduino automatically starts after the sketch is uploaded, if all is well, it will start a test sequence. Each LED in turn will cycle through the seven possible colors it can display: red, green, yellow, blue, mauve, cyan, and finally white.

If any of the LEDs do not light or do not show all seven colors, make a note of the problem and

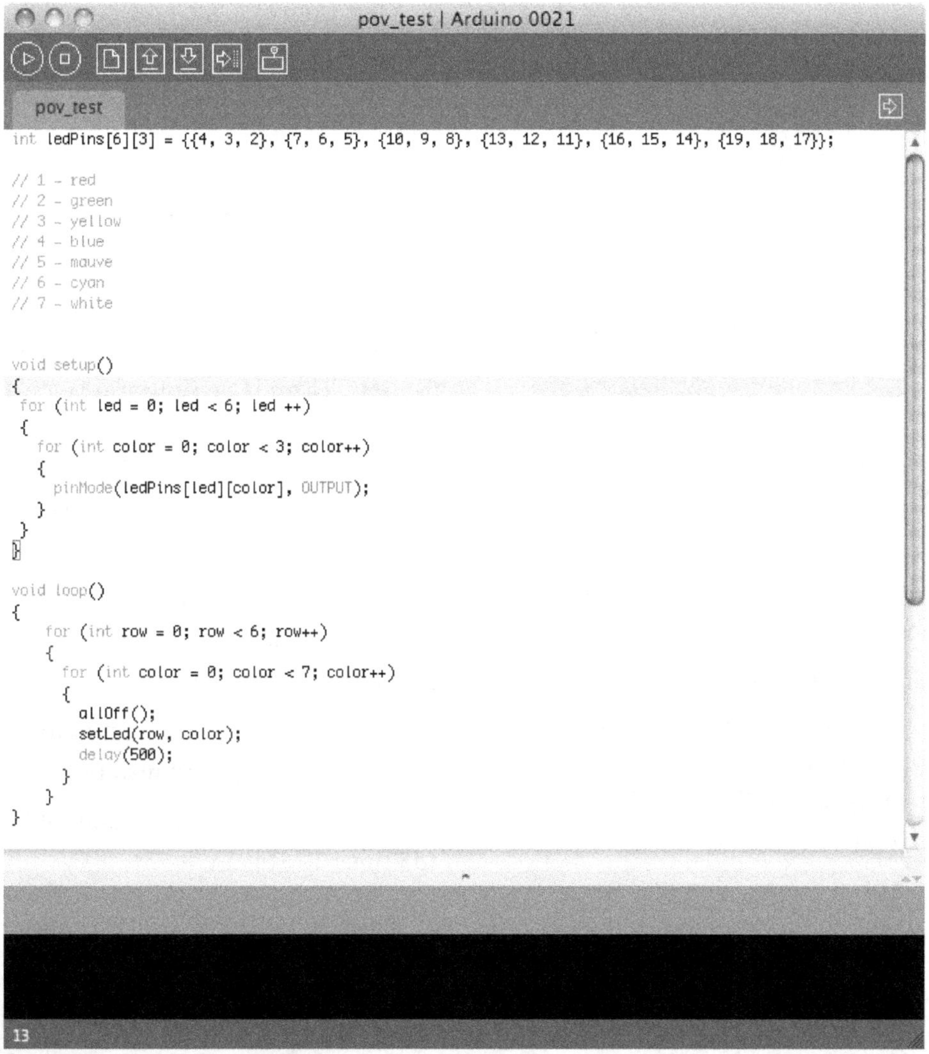

```
int ledPins[6][3] = {{4, 3, 2}, {7, 6, 5}, {10, 9, 8}, {13, 12, 11}, {16, 15, 14}, {19, 18, 17}};

// 1 - red
// 2 - green
// 3 - yellow
// 4 - blue
// 5 - mauve
// 6 - cyan
// 7 - white


void setup()
{
  for (int led = 0; led < 6; led ++)
  {
    for (int color = 0; color < 3; color++)
    {
      pinMode(ledPins[led][color], OUTPUT);
    }
  }
}

void loop()
{
  for (int row = 0; row < 6; row++)
  {
    for (int color = 0; color < 7; color++)
    {
      allOff();
      setLed(row, color);
      delay(500);
    }
  }
}
```

Figure 8-22 Loading the povtest sketch

refer back to the schematic diagram of Figure 8-3. Check your wiring and look for accidental solder bridges between tracks on the stripboard.

If the colors or the LED order seem all mixed up, you probably have one of the header pins to the Arduino the wrong way around.

Step 9. Make the Battery Connector

To power the project from a PP3 9V battery, we need to make ourselves a small lead that has a PP3 battery clip on one end and a 2.1mm power plug on the other. Figure 8-23 shows the semi-assembled lead. The red positive wire from the

battery lead is connected to the center connection of the plug and the back lead to the outside connection.

Keep the wires on the lead fairly short; otherwise, when the LED unit spins, the lead will fly out like a flail.

Step 10. Assemble the Woodwork

Well, woodwork rather overstates it. We just need a short strip of wood on which to mount the Arduino board and the LED stripboard. The length of the board is not critical, but it is a good idea to keep the overall weight of the module as low as

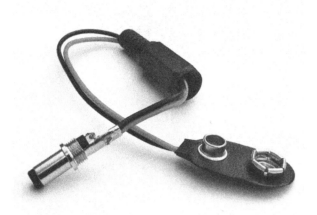

Figure 8-23   The battery clip for the LED unit

possible. The arrangement of the components on the strip of wood is shown in Figure 8-24.

The stripboard is screwed to one end of the wood. Be careful not to screw the screws in too tight, as the stripboard will crumble if compressed too hard. The rechargeable PP3 battery is fixed to the side of the wood using a cable tie.

Apart from attaching the assembly to a motor, which we will deal with in the next section, that's about it for the LED assembly.

Motor Controller

We need to be able to control the speed of the motor, because it must match the speed at which the LEDs switch from displaying one column to the next. There is no sensor to automatically synchronize the LEDs with the spinning motor. It's actually more fun without this, as the letters can be made to process around the display by tweaking the motor's speed.

You can buy readymade DC motor controllers, and if you do not want to make your own, this is probably the best option.

For those of you who want to make their own motor controller, please read on.

The motor controller uses our old friend, the 555 timer IC (Figure 8-25).

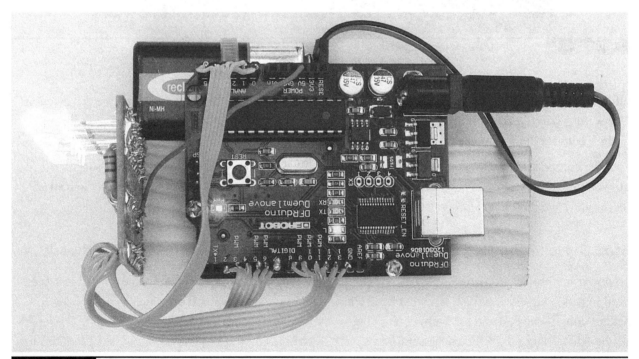

Figure 8-24   The components arranged on the wood

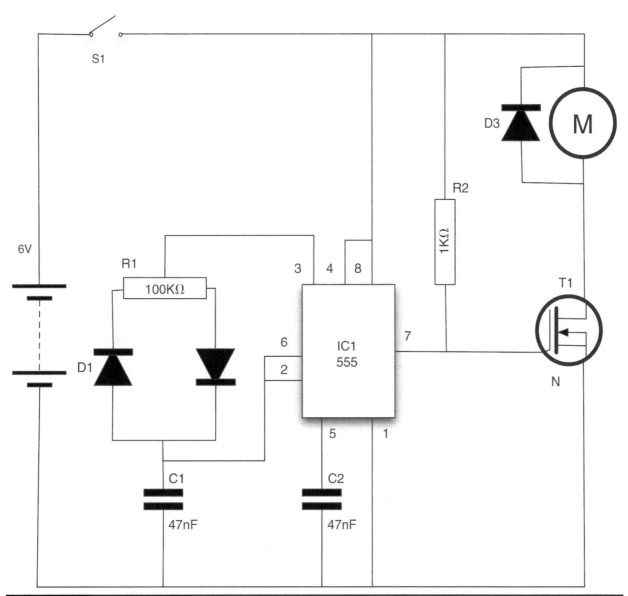

Figure 8-25 The schematic diagram of the motor controller

What You Will Need

To build the motor controller module, you will need the following components shown in the Parts Bin on the next page.

When sourcing a motor, try and find one with a mounting bracket, as this makes it much easier to fix it to the base. The motor the author used came in an educational kit.

Step 1. Prepare the Stripboard

Figure 8-26 shows the stripboard layout for the motor speed controller module.

The first step is to cut a piece of stripboard that has ten strips each of 20 holes. Six breaks must then be made in the tracks. Make these using a drill bit, which is used as a hand tool, rotating it between finger and thumb. Figure 8-27 shows the prepared board ready for the components to be soldered to it.

PARTS BIN			
Part	**Quantity**	**Description**	**Source**
R1	1	100kΩ linear potentiometer	Farnell: 1227589
R2	1	1kΩ 0.5-W metal film resistor	Farnell: 9339779
D1-3	3	1N4004 diode	Farnell: 9109595
C1, C2	2	47nF capacitor	Farnell: 1200415
T1	1	N-Channel MOSFET	Farnell: 9845534
S1	1	SPST miniature toggle switch	Farnell: 1661841
Battery clip	1	PP3 battery clip	Farnell: 1650667
Battery holder	1	Holder for four AA cells	Farnell: 1696782
Battery	4	1.5V or rechargeable AA cells	
Stripboard	1	Stripboard; ten tracks, each with 20 holes	Farnell: 1201473
Terminal block	1	Two-way terminal block	Farnell: 1055837
Knob	1	Small plastic knob	Farnell: 1441137
Motor	1	6V DC motor	
Box	1	Plastic project box 4½" × 3" (120mm × 80mm)	

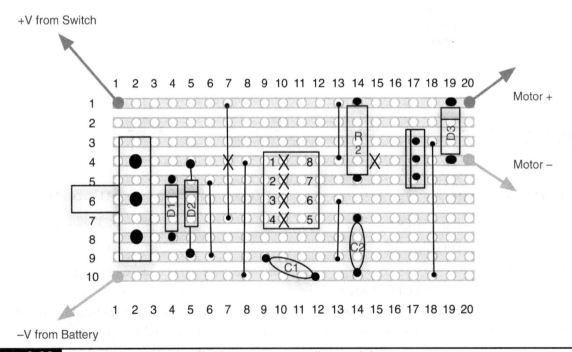

Figure 8-26 The stripboard layout for the motor controller module

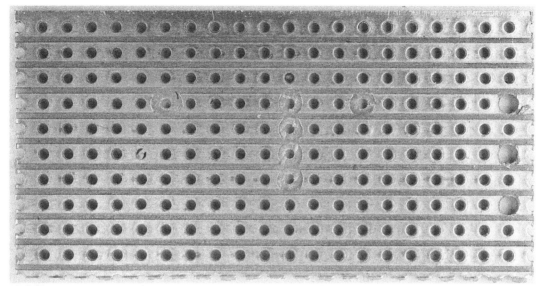

Figure 8-27   The motor controller module stripboard ready for soldering

The three holes for the variable resistor may need enlarging with a drill.

Step 2. Solder the Links Resistor and Diodes

Before starting on the real components, solder the three wire links in place and then the resistors. Once done, your board should look like that in Figure 8-28.

Step 3. Solder the Remaining Components

We can now solder the rest of the components, starting with those lowest in profile. Solder the IC first, and then the capacitors and transistor.

Finally, fit the variable resistor. If enlarging the holes in the stripboard for the variable resistor has removed some of the track, then after putting the variable resistor in place, bend the leads over

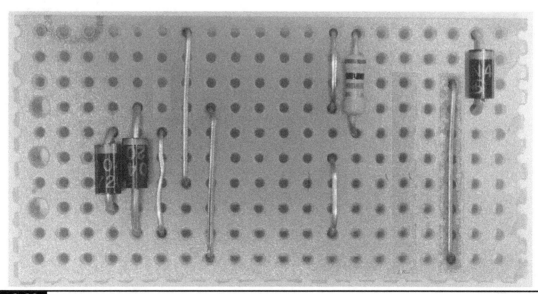

Figure 8-28   The board with resistor, diodes, and links

toward the center of the board so they can be soldered over a few holes worth of track.

We are also going to attach a screw terminal for connecting the motor leads. So, solder a pair of short half-inch (10mm) wires to the appropriate holes (see Figure 8-26). You should find that some of the snipped resistor leads should be about the right length.

The completed board is shown in Figure 8-29.

Step 4. Wire Up the Motor Controller

This step is straightforward, because the variable resistor is already mounted on the stripboard, so the only things that need wiring up are the switch, the battery clip, and the motor itself.

Figure 8-30 shows how the motor controller is wired up.

Platform

We now need to construct a base containing the motor and a motor controller to drive the small DC motor that will turn our LED module. The motor

will vibrate when the LED arm is attached, so it is a good idea to build a solid base for the project.

The base the author used takes this to an extreme and is constructed from reclaimed two-by-four wood, crudely screwed together (Figure 8-31).

The dimensions of this are really not critical at all, as long as the motor can be mounted in such a way that its spindle is not impeded and there is access to the terminals of the motor to solder leads to them.

Putting It All Together

We can now test the motor controller before we fit it into a box. Attach the batteries and the motor. Turn the switch on and the motor should whir. The variable resistor should allow the speed of the motor to be altered from stationary to fast. Leave the knob at the slow end for now.

If everything is okay, we can fit the motor controller and batteries into a box (Figure 8-32). Holes will need to be drilled for the variable resistor and switch. It is a good idea to lay all the parts inside the box and make sure they all fit before you start drilling holes.

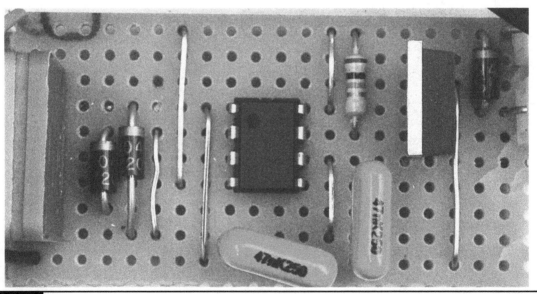

Figure 8-29   The completed motor controller stripboard

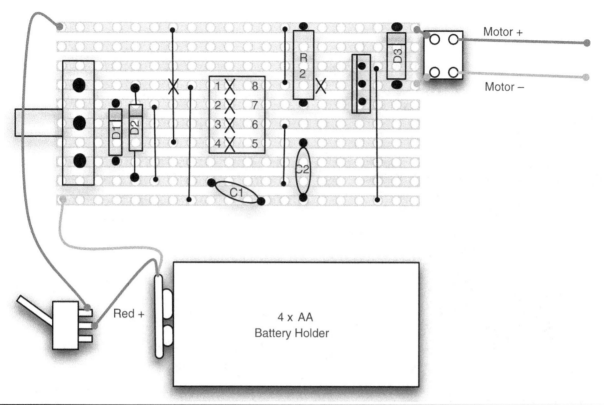

Figure 8-30 The motor controller wiring diagram

We are going to need a means of attaching a motor spindle to the bottom of the wood. This is quite tricky because it needs to keep the wood as flat as possible while giving a tight connection to the motor spindle. You will need your motor on hand to try out the size.

The author used the spindle from a broken CD drive. This had four screw holes that could be attached to the wooden arm, as well as a central hole that was a little too small for the motor spindle but that was carefully drilled to the correct size for a tight fit. (See Figure 8-33.)

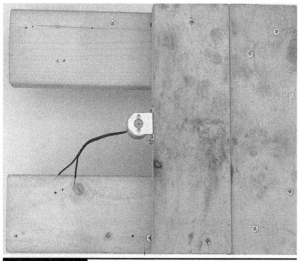

Figure 8-31 A wooden base for the motor

Figure 8-32 The motor controller and batteries boxed

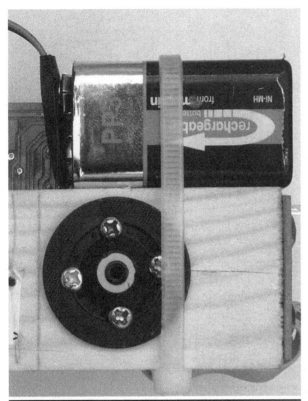

Figure 8-33 Attaching the arm to the motor

This is where, as Evil Geniuses, you must use your inventiveness and scour your junk boxes for a means of attaching your motor.

When attaching the motor, it is important for the spindle to enter the arm as close to the arm's center of gravity as possible to minimize vibrations. You can determine this point by finding the line along the length of the arm at which it just balances. Balancing the arm on a nail works. The spindle should also be perpendicular to the wooden arm, otherwise the whole arm will vibrate up and down, spoiling the persistence-of-vision effect.

Before we turn it on, we need to install the real sketch onto the Arduino, rather than the sketch we were using to test the LEDs. So, go to www.dangerouslymad.com and copy the pov script into a new Arduino project and upload it to the board.

That's it. Time to turn it on and try it out! Connect the battery of the arm to the socket of the Arduino board, and after a moment the LEDs

should begin to flash. Now turn on the motor and slowly increase the speed. If things start to feel unstable, turn everything off and try balancing the arm better or use a different means of connecting the arm to the spindle. Be careful at this stage, because the vibrations may loosen the arm, causing it to fly off and potentially break or hurt someone.

You will start to pick out some fast-moving letters. Concentrate on one of the letters and adjust the speed of the motor until the letters stands still. You should then be able to set the letters to appear to progress around the cylinder of light by adjusting the speed slightly.

Changing the Message

In the "Theory" section later in this chapter, we explain just how the software works, but for now you are probably wondering how to change the message that is displayed. The only way to do this is by modifying the sketch and uploading it to the Arduino board again.

The relevant part of the sketch is shown highlighted in Figure 8-34.

If you look closely, you can see the letters HELLO in the program. Each number represents a color. The number 0 means completely off, 1 means red, and so on, as shown in the comments immediately above the highlighted area.

The line that says:

```
int n = 64;
```

specifies how many columns the message is in length. So you can construct a longer or shorter message than "hello," but you will need to alter the value of "n" accordingly.

So, to change the message, just change the numbers in the highlighted area. This is quite a tedious process, so to make it easier the author has devised a web-based tool to help you convert text into the appropriate array of numbers. You

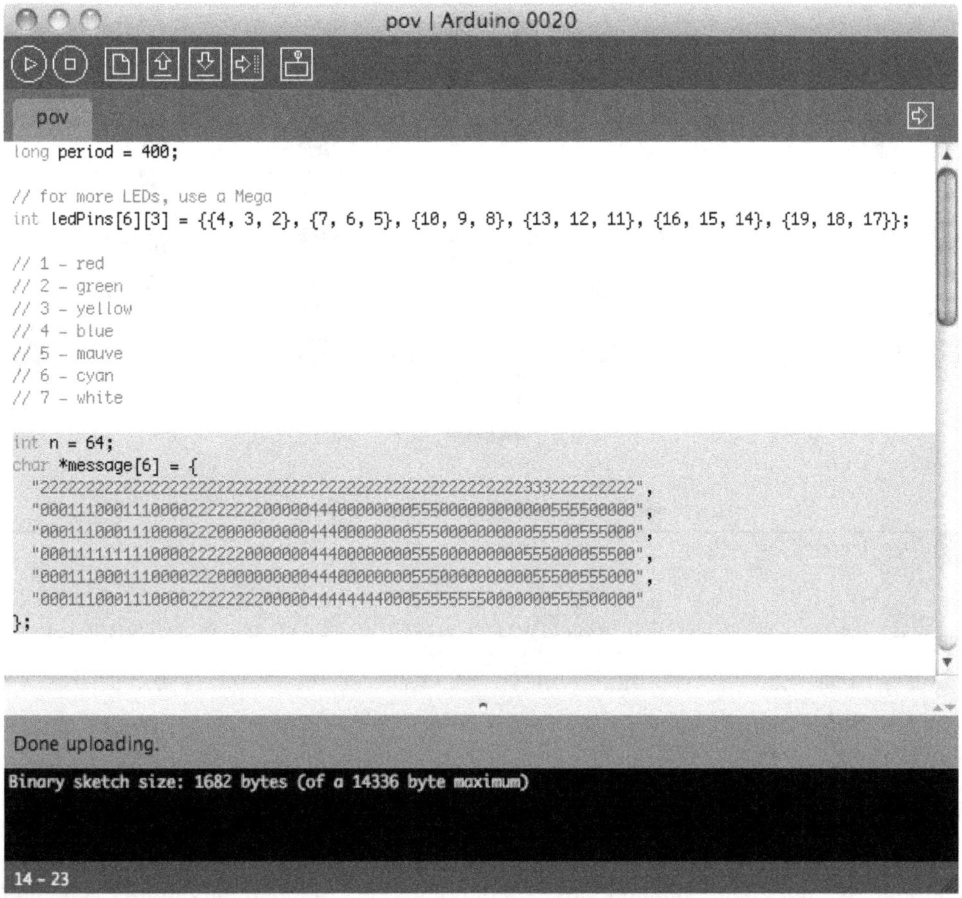

```
long period = 400;

// for more LEDs, use a Mega
int ledPins[6][3] = {{4, 3, 2}, {7, 6, 5}, {10, 9, 8}, {13, 12, 11}, {16, 15, 14}, {19, 18, 17}};

// 1 - red
// 2 - green
// 3 - yellow
// 4 - blue
// 5 - mauve
// 6 - cyan
// 7 - white

int n = 64;
char *message[6] = {
  "2222222222222222222222222222222222222222222222222222333222222222",
  "0001110001110000222222220000044400000000555000000000000555500000",
  "0001110001110000222000000000044400000000555000000000055500555000",
  "0001111111110000222220000000044000000000555000000000555000055500",
  "0001110001110000222200000000044400000000555000000000055500555000",
  "0001110001110000222222220000044444444000555555555000000555500000"
};
```

Done uploading.

Binary sketch size: 1682 bytes (of a 14336 byte maximum)

14 - 23

Figure 8-34　Modifying the message

can find this tool on the book's web site at www.dangerouslymad.com.

Theory

This project includes some interesting bits of theory. There is the psychological persistence-of-vision effect, and the software to achieve the effect, as well as the use of pulse width modulation (PWM) to control the speed of a motor.

POV

This project works for the same reason that movies work. If images are presented to the brain faster than around 25 frames per second, then we interpret the image as moving. In this case, we are not presenting a whole image at once, but rather

constructing an image by turning LEDs on and off as they move past the observer's field of view.

Figure 8-35 illustrates this point.

The Software

The Arduino sketch for the Persistence of Vision project in shown in Listing 8-1 on the next page.

The first line of the program, or "sketch" as they are known in the Arduino community, sets the "period" in microseconds between displaying each successive column. You can experiment with changing this value.

We then define a two-dimensional array to represent the Arduino pins used by the LEDs. The first dimension is the LED's row and the second is the individual red, green, or blue element within the LED. So, as we know from the wiring diagram,

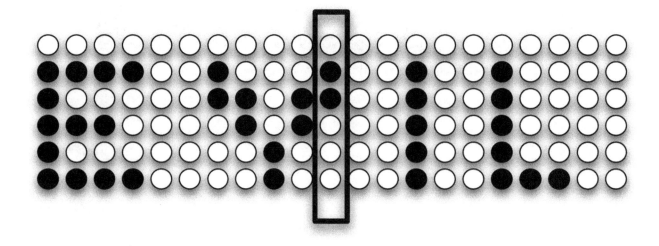

Figure 8-35   Persistence of vision

LISTING 8-1

```
long period = 400;
int ledPins[6][3] = {{4, 3, 2}, {7, 6, 5}, {10, 9, 8}, {13, 12, 11},
    {16, 15, 14}, {19, 18, 17}};
int n = 128;
char *message[6] = {
"000000000000000000000000000000000000000000000000000000000000000000000000000000
    0000000000000000000000000000000000770000000000000",
"111111100220000022000333300044000000000000000000005555550066666660011000110002222000
    0330003300044444000000000000000000770000000000000",
"110000000022000220000033000044000000077770000005500000066000000001110011000022000
    0330003300044000000000000000000000770000000000000",
"111110000000220220000000330000440000000777700000055555500066666000011110110000220000
    0330003300044444000000000000000000770000000000000",
"110000000000022200000000330000440000000000000000005500055006600000000110011100002200
    0330003300000004400000000000000000770000000000000",
"111111100000002000000033330004444400000000000000005555500066666660011000110002222200
    00333333000444444000000000000000000770000000000000"
};
void setup()
{
  for (int led = 0; led < 6; led ++)
  {
    for (int color = 0; color < 3; color++)
    {
      pinMode(ledPins[led][color], OUTPUT);
    }
  }
}
void loop()
```

LISTING 8-1   *(continued)*

```
{
   for (int col = 0; col < n; col++)
   {
      for (int row = 0; row < 6; row++)
      {
         int color = (int)(message[5-row][col] - '0');
         setLed(row, color);
      }
      delayMicroseconds(period);
      allOff();
      delayMicroseconds(period / 16);
   }
}
void setLed(int led, int color)
{
 digitalWrite(ledPins[led][0], !(color & 1));
 digitalWrite(ledPins[led][1], !(color & 2));
 digitalWrite(ledPins[led][2], !(color & 4));
}
void allOff()
{
   for (int led = 0; led < 6; led ++)
 {
   for (int color = 0; color < 3; color++)
   {
      digitalWrite(ledPins[led][color], HIGH);
   }
 }
}
```

the red LED of the first row is connected to Arduino pin 4, the green to pin 3, and the blue to pin 2.

Next, we have the data for the message to be displayed. If you look carefully, you can see the message in the numbers rather like the film *The Matrix*, only not quite as cool. Each number represents a color.

The "setup" function simply initializes all the LEDs to be output pins. This function is called automatically just once, when the Arduino board starts up.

The "loop" function is called continuously and it is here that we put the code to switch between the columns. It contains two nested loops. The outer loop steps through each column in turn, then inner loop each LED. At the center of the loop, it then sets the appropriate LED color using the setLed function.

After the LEDs for each column have been displayed, it pauses for a time specified in the period variable and then turns off all the LEDs using the allOff function before pausing again for $\frac{1}{16}$ of the period. This leaves a short gap between the columns, making the display more readable.

The setLed function sets the three colors of a particular LED on or off depending on the color number between 1 and 7. The number between 0 and 7 is a three-bit number (three binary digits). Each of these digits can be 1 or 0, which is on or off. So a number 7 in decimal is actually 111 in binary, which means that all three colors would be on, making the LED appear white.

The expression:

```
!(color & 2)
```

will evaluate to 0 if the second bit of the color number is set (green), and 2 if it is not. This may seem the wrong way around, but the LEDs are common anode, which means they turn on when the output pins are at 0V rather than at 5V.

The allOff function simply iterates over every LED pin, setting it to 5V.

Motor Controller

PWM stands for pulse width modulation, and refers to the means of controlling the amount of power delivered to a motor in this case. We use a simple PWM controller built using a 555 timer IC. We will meet the 555 timer again in Chapter 12.

PWM controls the power to a motor by lengthening or shortening pulses of power that arrive at the motor. You could imagine this as a child sitting on a swing. If you just give the child short pushes, they will not swing as high as if you give the child a longer push.

In a motor, the motor either receives full power or no power, in pulses that arrive at perhaps a thousand times per second. The motor has no time to fully stop when the power is off, so the overall effect is that the speed of the motor is controlled smoothly.

Summary

This project could be scaled up to use a bigger number of LEDs, although you would probably have to use the Arduino's bigger brother, the Arduino Mega, as this has a lot more input/output pins on it.

If you search the Internet for "persistence of vision," you will find many interesting ideas. People have even displayed video using this approach.

In the next chapter, we switch our attention to spying.

CHAPTER 9

Covert Radio Bug

PROJECT SIZE: Medium

SKILL LEVEL: ★★★☆

THE EVIL GENIUS LIKES TO KNOW what his enemies are saying. And what better way to discover that than to plant a bug. On the other hand, the Evil Genius considers any spying on *him* as an unacceptable breach of his privacy, and for this reason, this chapter also includes a "bug detector" that will let him locate bugs that may have found their way into his lair. (Figure 9-1 shows the bug and Figure 9-2 the bug detector.)

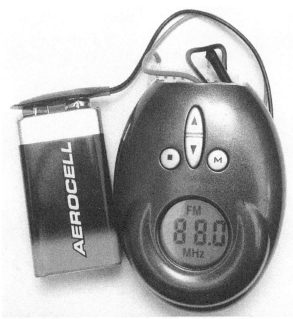

Figure 9-1 The FM transmitter

Figure 9-2 The bug detector

Rather than build a bug from scratch, this project cunningly uses a cheap FM transmitter of the type employed in cars to play music from an MP3 player. These devices can be acquired very cheaply and can be set to transmit at any frequency on the FM band. Ideally, this should be at a frequency not in use by a radio station, but the Evil Genius will occasionally confuse his minions by deliberately setting it to the frequency of their favorite station and barking commands at them over the transmitter.

The shocked looks on their little faces when they hear their radio addressing them personally, and in such aggressive terms, is priceless to the Evil Genius.

The Bug

FM transmitters for MP3 players expect to receive a signal from an MP3 player that is already amplified to a reasonable level. Unfortunately, microphones only produce a small electrical signal that must be amplified before anything useful can be done with it. So we have to make a little amplifier to feed the transmitter. Figure 9-3 shows the schematic diagram for the bug.

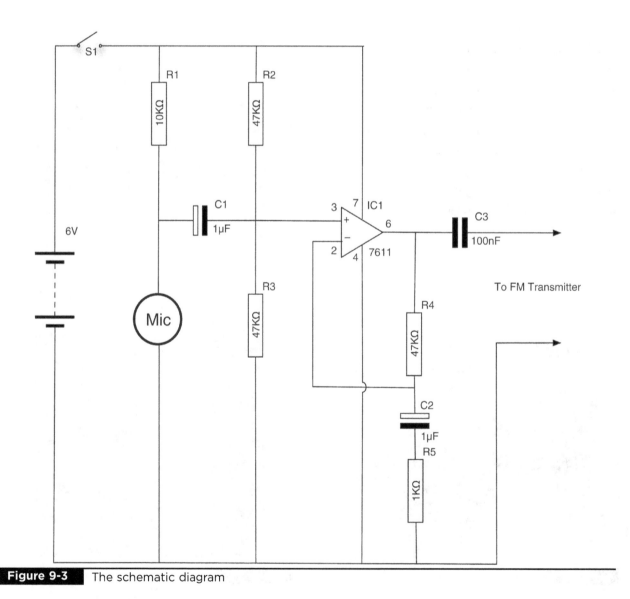

Figure 9-3   The schematic diagram

What You Will Need

You will need the components listed in the Parts Bin to build the bug.

You will also need the following tools:

T O O L B O X
▪ Soldering equipment
▪ A hot glue gun or self-adhesive pads
▪ A hacksaw

Assembly

The project is all assembled on a single piece of stripboard. The following steps will lead you through the construction.

Step 1. Prepare the Stripboard

Figure 9-4 shows the stripboard layout for the microphone amplifier.

The first step is to cut a piece of stripboard that has ten strips, each with 22 holes. Seven breaks should be made in the tracks. Make these using a drill bit, employing the bit as a hand tool by rotating it between your finger and thumb. Figure 9-5 shows the prepared board ready for the components to be soldered onto it.

Step 2. Solder the Links

Before starting on the real components, solder the four wire links into place. When this is complete, your board should look like Figure 9-6.

PARTS BIN			
Part	**Quantity**	**Description**	**Source**
Transmitter	1	FM transmitter for MP3 player	eBay
R1	1	10kΩ 0.5-W metal film resistor	Farnell: 9339787
R2–R4	3	47kΩ 0.5-W metal film resistor	Farnell: 9340637
R5	1	1kΩ 0.5-W metal film resistor	Farnell: 9339779
C1, C2	2	1μF 16V electrolytic capacitor	Farnell: 1236655
C3	1	100nF	Farnell: 1200414
IC1	1	7611 op amp	Farnell: 1018166
IC socket	1	8 pin DIL IC socket	Farnell: 1101345
Mic	1	Electret microphone insert	Farnell: 1736563
Switch	1	SPST toggle switch	Farnell: 1661841
Battery clip	1	PP3 battery clip	Farnell: 1183124
Stripboard		Ten strips each with 22 holes	Farnell: 1201473
Battery	1	9V PP3 battery	

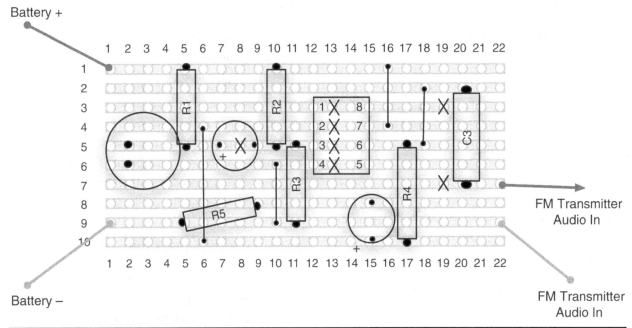

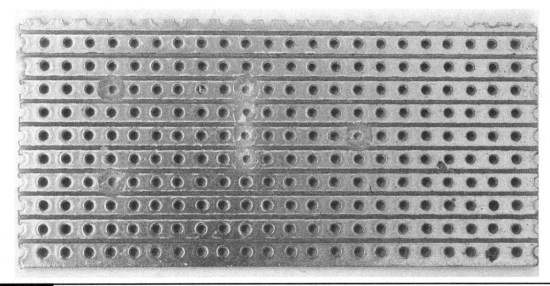

Figure 9-5   The stripboard for the bug

Step 3. Solder the Resistors

We can now solder all the resistors in place. Take special care with R5, which is soldered at an angle across one of the links. The leads of the resistor should be kept away from the wire of the link. However, it does not matter if the body of the resistor touches the link.

When all the resistors are soldered into place, your board will look like Figure 9-7.

Step 4. Solder the IC

Place the IC in position (see Figure 9-8), ensuring that it is the right way around—the little dot indicating pin 1 should be on the top left. Solder

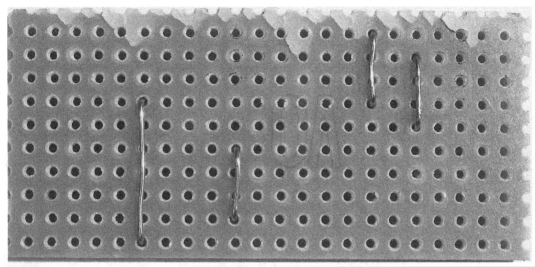

Figure 9-6 The bug stripboard with links

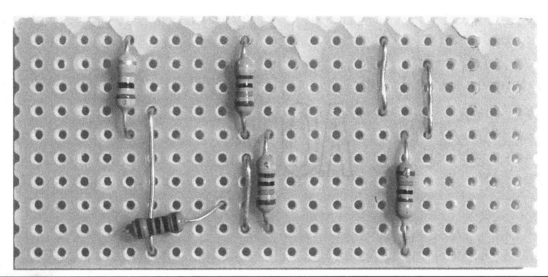

Figure 9-7 The bug board with resistors

the IC into place, using the same caution as with the receiver's construction. Like with the receiver, you may decide to use an IC socket rather than solder the IC directly.

Step 5. Solder the Remaining Components

The rest of the components can now be soldered onto the board. Solder the transistor first (making sure it is the right way around), followed by the capacitors and the electret microphone insert.

The microphone that the author used was designed to be soldered to a PCB. However, you may have to solder short leads to the contacts on the microphone. Position the microphone as shown in Figure 9-8 so it is pointing away from the rest of the electronics.

The microphone also must be connected the correct way around. The contact that is connected to the case should be the negative lead, which should be the lower contact on the stripboard. You may need to test this with a multimeter. If the

Figure 9-8   The completed transmitter stripboard

transmitter does not work, try the microphone the other way around.

Step 6. Prepare the Transmitter

Figure 9-9 shows the FM transmitter in its unmodified state. These things are designed to fit into your car's cigarette lighter socket and have a trailing lead that plugs into your MP3 player. The unit has a screen and some buttons that allow you to select the frequency on which it broadcasts. You would then tune your car radio to that frequency to be able to hear your MP3 player through the car's speaker system.

These devices can be bought very cheaply, and many types are available on eBay for a few dollars.

Although designed to operate from the car's 12V power supply, these units normally contain a voltage regulator that drops the voltage to 5V. This means they are equally happy operating from a 9V PP3 battery.

If you take your FM transmitter apart, it should look something like Figure 9-10.

On the left-hand side, you can see the power leads connected to the car accessory plug. These can be cut off, leaving leads of two or three inches

that can be stripped and tinned, and made ready for soldering. This unit also had wires that lead to a backup battery (toward the bottom of Figure 9-10), a setup that was designed to remember the radio's frequency setting in between uses of the device. This feature was deemed an unnecessary complication and those leads can be cut off. Finally, the audio lead can be shortened, stripped, and tinned, leaving a raw module like that in Figure 9-11.

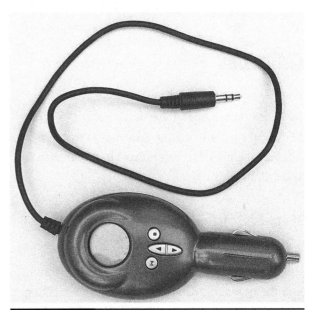

Figure 9-9   FM MP3 car transmitter

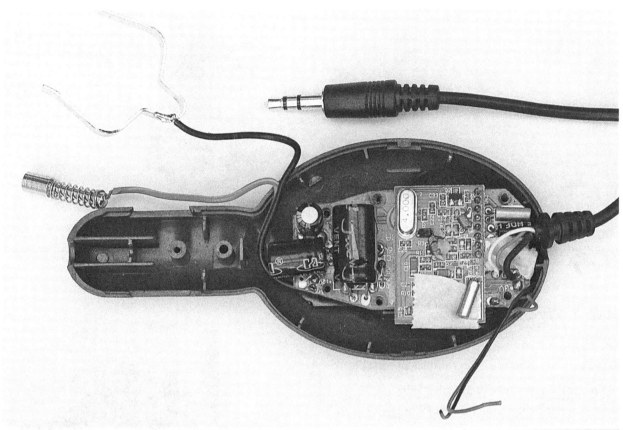

Figure 9-10 FM transmitter disassembled

Figure 9-11 FM transmitter with leads trimmed

The design of this unit used an LCD screen that was simply pressed against a connector leading to the circuit board. This required the board to be attached to the case, to hold the display in place. If your unit's screen is attached more strongly, you can leave the case off entirely. In this instance, we attached the case but cut off the long plug part with a hacksaw (Figure 9-12).

In Figure 9-12, you can also make out crocodile clips that were connected to a 9V supply to make sure our transmitter wasn't damaged during its surgery. The audio lead has also been stripped and tinned. Note that the left and right internal wires have been soldered together. Our bug will not be in stereo.

Figure 9-12 FM transmitter reduced to basics

Step 7. Wire Everything Together

Referring back to Figure 9-4, wire up the project so we can test it before fitting it into a project box.

Solder the red lead of the battery clip to the stripboard, and then solder the black negative lead from the battery clip to the negative connection on the stripboard.

Next, attach the power leads of the FM transmitter to the stripboard and finally the audio input of the transmitter to the stripboard. The shielding of the audio lead does not need connecting as it will already be connected to ground. Both the left and right audio wires inside the shielding should be connected together before attaching them to the stripboard (Figure 9-13).

Step 8. Test the Bug

Attach the battery and set the FM transmitter to an unused frequency. Place it near the TV or another sound source, and retire to a neighboring room with a portable FM radio tuned to the same frequency.

You should be able to hear the sound coming from the bug.

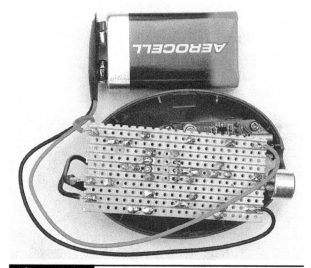

Figure 9-13 The bug assembled

Using the Bug

Unlike you, the reader, the Evil Genius has no qualms about listening to other people's conversations. He will conceal his bug in a small box—perhaps a pack of cards—and retire to another room with his transistor radio and a flask of warm milk. His minions have learned that their master likes to operate in this fashion, and so they *always* say nice things about him. This pleases the Evil Genius.

A Bug Detector

One of the interesting things about this bug is that because of its short range and low power, it broadcasts at a strength that is no stronger than the radio signals received from commercial radio stations in the FM band. This makes it almost impossible to build a detector to find out if you are being bugged using this kind of device.

A lot of modern bugs are based on cell phone technology. They are basically a phone that can be concealed somewhere and turned on remotely either by a phone call or SMS message. It is much easier to detect such bugs—at least when they are on.

So, we are going to build a detector for these kinds of bugs. The Evil Genius also finds it a good way to detect minions using their cell phones when they should be doing their minion homework. This particularly annoys the Evil Genius, especially when their phones are better than his. So having detected such errant behavior, the Evil Genius can enjoy a good cell phone stomping session.

The detector looks like a detector should look (Figure 9-14), with a proper analog meter, whose needle swings over to indicate the strength of the microwave frequencies that cell phones use.

The antenna is attached using screw terminals to allow easy switching between different antenna designs. This will allow you to experiment with different designs to work with the different frequencies at which mobile phones operate.

Figure 9-15 shows the schematic diagram for the bug detector.

The design uses a germanium diode as a detector and amplifies the weak signal from the phone using an operational amplifier. For a more detailed description, see the "Theory" section that follows.

Figure 9-14 Cell phone bug detector

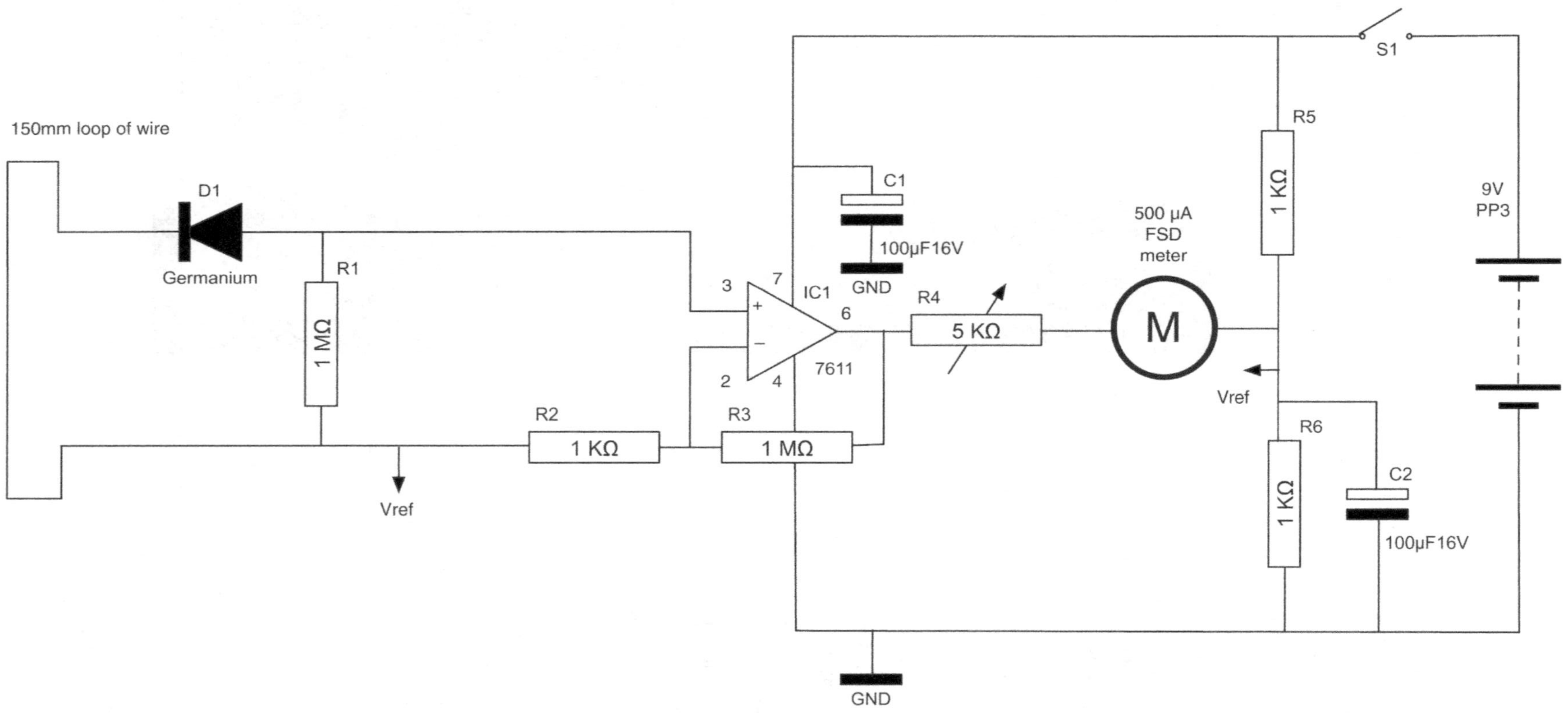

Figure 9-15 Schematic diagram for the cell phone detector

What You Will Need

You will need the components listed in the Parts Bin to build the cell phone detector.

You will also need the following tools:

TOOLBOX
▪ Soldering equipment
▪ A drill and assorted drill bits
▪ A hot glue gun or self-adhesive pads

Assembly

The bug detector is built onto a small piece of stripboard, with the meter variable resistor and switch mounted in the lid of the plastic case. The instructions below will lead you step by step through the process of building the detector.

Step 1. Prepare the Stripboard

Figure 9-16 shows the stripboard layout for the cell phone detector.

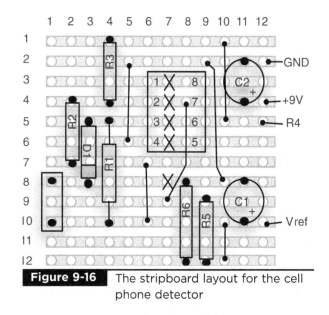

Figure 9-16 The stripboard layout for the cell phone detector

PARTS BIN			
Part	**Quantity**	**Description**	**Source**
R1, R3	2	1MΩ 0.5-W metal film resistor	Farnell: 9339809
R2, R5, R6	3	1kΩ 0.5-W metal film resistor	Farnell: 9339779
R4	1	5kΩ variable resistor (potentiometer)	Farnell: 1417135
C1, C2	2	100µF 16V electrolytic capacitor	Farnell: 1136275
D1	1	Germanium diode – OA91	Farnell: 1208635
IC1	1	7611 op amp	Farnell: 1018166
IC socket (optional)	1	Eight-pin DIL IC socket	Farnell: 1101345
Meter	1	500µA FSD meter	Farnell: 7758219
Switch	1	SPST toggle switch	Farnell: 1661841
Battery clip	1	PP3 battery clip	Farnell: 1183124
Stripboard		12 strips each of 12 holes	Farnell: 1201473
Terminal block	1		Farnell: 1055837
Wire		Thick solid core wire for antenna (20 AWG)	
Battery	1	9V PP3 battery	

The first step is to cut a piece of stripboard that has 12 strips, each with 12 holes. Five breaks must then be made in the tracks. Make these using a drill bit—that is, just using it as a hand tool and rotating it between your finger and thumb. Figure 9-17 shows the prepared board ready to have the components soldered to it.

Step 2. Solder the Links and Resistors

Before starting on the real components, solder the six wire links into place (Figure 9-18). Note that you will need to bend some of the links to fit them all on the board.

We can now solder all the resistors in place and also solder in the diode. Be careful to get the diode the correct way around.

When all the resistors and the diode are soldered into place, your board will look like Figure 9-19.

Step 3. Solder the IC

Solder the IC next. Make sure that pin 1 is at the top of the board. Pin 1 is marked by a little circle

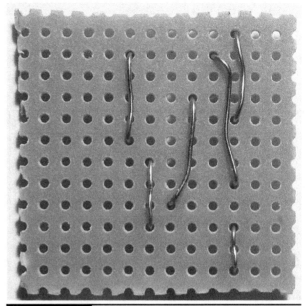

Figure 9-18   The board with links

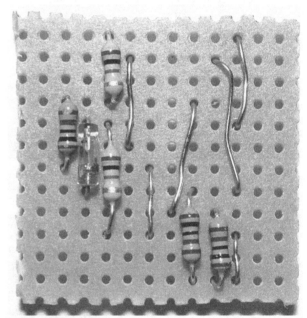

Figure 9-19   The board with resistors and diode

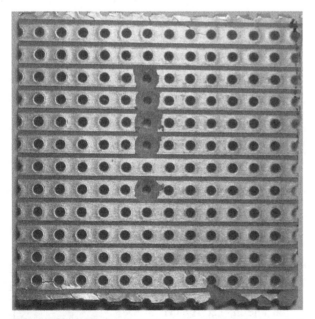

Figure 9-17   The cell phone detector stripboard ready for soldering

on the IC package. If you are worried about soldering the IC directly onto the stripboard, you can use an eight-pin IC socket.

Note how the IC sits on top of one of the links. Make sure that none of the links are touching the IC pins.

When the IC is in place, the board should look like Figure 9-20.

Figure 9-20 The board with the IC in place

Figure 9-21 The board with all the components in place

Step 4. Solder the Capacitors and Terminal Block

We can now solder in the terminal block and the capacitors. Make sure the capacitors are the correct way around. They should both have the longer positive lead toward the bottom of the stripboard.

The board with all the components in place is shown in Figure 9-21.

Step 5. Make an Antenna

The antenna is just a loop of wire 150mm in length. The author used 20-AWG enamel-coated wire. The thickness is not critical, but the wire should be thick enough to hold its shape. If you use enameled wire, you will have to strip off the enamel at the wire ends. Use a pair of snips or a knife to do this before inserting it into the terminal block (Figure 9-22).

Step 6. Test the Bug Detector

Before fixing everything into a box, we need to test out the circuit and make sure everything works

as it should. So, connect up the components as shown in the wiring diagram of Figure 9-23.

Start by soldering short leads to the center and one end of the variable resistor. One of these leads will be connected to the negative terminal of the meter. Another short lead should be soldered from the positive lead of the meter to the Vref connection track on the stripboard next to C1.

The other lead from the variable resistor should be soldered to the R4 strip on the breadboard.

We now just need to connect up the switch and battery clip. Cut the positive (red) lead of the battery clip about halfway along its length and solder in the switch, as shown in Figure 9-23. Then, solder the remainder of the positive battery clip lead between the switch and the +9V strip on the stripboard.

Turn the variable resistor to its halfway position, attach the battery, and turn on the switch. The meter should move slightly away from its zero position. If it does not move, or the meter pings hard over to the far side, then turn off the power immediately and carefully check your circuit.

Now get a mobile phone and, holding it close to the antenna, make a call to your voicemail. You

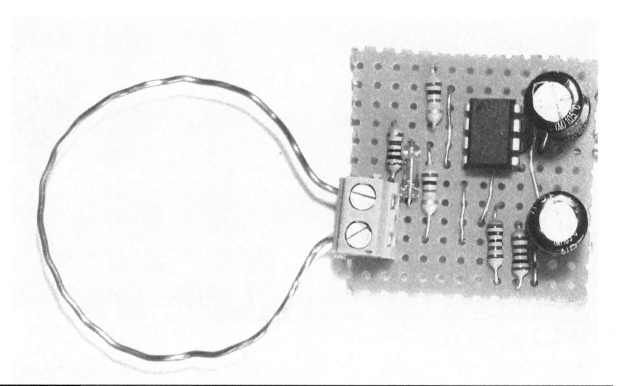

Figure 9-22   The antenna attached to the stripboard

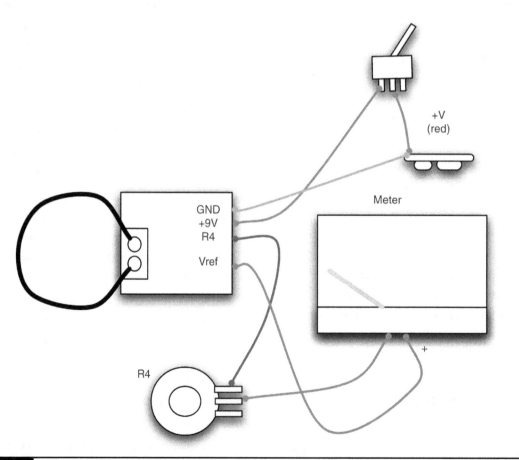

Figure 9-23   Wiring diagram for the cell phone detector

should see the meter start to twitch and then move quite vigorously. You'll find that the meter is sensitive enough to detect a phone call from a few meters away.

Step 7. Box the Project

Assuming everything is working, we can put the project in a box if we like. As with all the projects in this book, first find a box that is big enough, and then work out where the project parts will fit before drilling holes for the switch variable resistor and the meter.

Figure 9-24 shows the layout of the components in the box.

Fix the stripboard in place at the end of the box, with the terminal block pressed against the end of the box. Use self-adhesive pads or blobs of glue

from a hot glue gun. Drill two small holes for the wire ends of the antenna. This is going to be outside of the box, so the holes need to line up with the terminal block.

Theory

Referring back to Figure 9-15, a small signal in the gigahertz frequency will be induced in the loop antenna. This is rectified (see Figure 9-25) by the germanium diode. A germanium diode is used because it has a much lower voltage drop than the usual silicon diode (0.3V compared to 0.6V). This makes it more sensitive.

To measure how strong the signal is, think about the average signal. If it is not rectified, then half the time it is positive and half the time it is negative, making the average signal always zero.

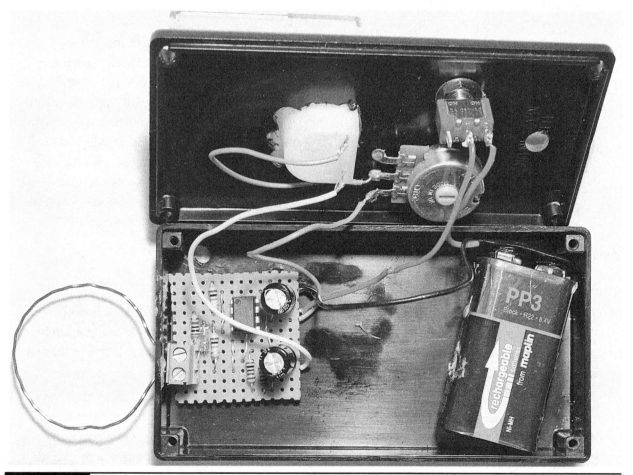

Figure 9-24   Fitting the cell phone detector into a box

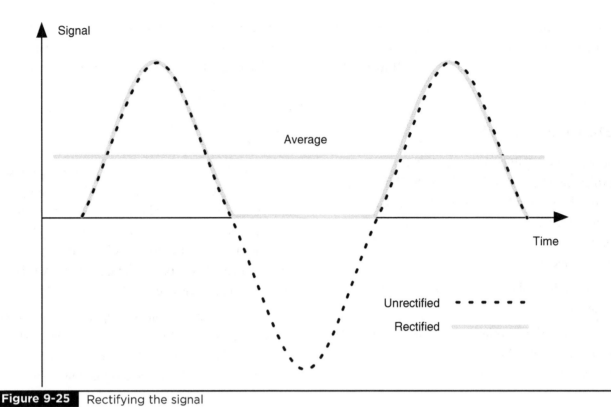

Figure 9-25   Rectifying the signal

By rectifying it, we chop off the negative bit, making our average signal positive in proportion to how big the waves are.

This signal is oscillating at over a billion times a second, so we do not need to consider smoothing it with a capacitor since the rest of the electronics won't be fast enough to see anything more than an average anyway.

The next step is to amplify that weak signal and we use an amplifier IC to do this. The type of amplifier we use is called an operational amplifier, or "op amp" for short. This kind of amplifier has two inputs, one positive and one negative. It amplifies the difference between these two inputs. However, to make the amplifier stable, you always operate it with feedback. This passes a portion of the amplified signal back to the input, letting you control exactly how much it amplifies (its gain).

In the arrangement we've used here, which is called "non-inverting," we employ the negative input just for feedback and pass the signal we want to amplify into the positive input.

The output will be the input multiplied by the gain. The gain is set by the degree of negative feedback and can be calculated as $1 + R3 / R2$. Since R3 is 1MΩ, and R2 is 1K, the gain is 1001. Op amps are really designed to be operated with a split supply. For example, +9V, GND, and –9V. To do this would require two batteries, which we do not really want. So, instead we use the arrangement of R5, R6, and C2 to provide a reference voltage of about 4.5V, halfway between GND and 9V.

Summary

Now the Evil Genius has a means of bugging people, and also a means of detecting when he's been bugged (albeit only if the bug is of the cell phone variety).

In the next project, we will make a system for sending sound over a laser beam. No radio waves means no way of eavesdropping with a radio receiver, and so the Evil Genius can once again communicate securely.

CHAPTER 10

Laser Voice Transmitter

PROJECT SIZE: Large

SKILL LEVEL: ★★★☆

THE EVIL GENIUS LIKES to communicate with his friend (also an Evil Genius) who lives in the house opposite. Radio waves are far too easily intercepted, and Morse code with a flashlight is just plain tricky. So, the Evil Genius uses a laser beam that passes through the glass of his window, across the road, and through the glass of his friend's window, where it hits a receiver so that the friend can hear what he is saying (see Figure 10-1).

Not being terribly interested in what his friend has to say in return, he is content to watch her nod, indicating she has heard the wise words of the Evil Genius.

One day, the Evil Genius will make another receiver and transmitter pair so his friend can respond with something more than just gestures.

The project is comprised of two parts: the receiver, which amplifies the signal from the laser and plays it on the speaker; and the transmitter that modulates the laser beam with the sound wave.

The Receiver

The receiver is the simplest part of the project. It uses an IC audio amplifier to directly amplify the

signal from a phototransistor. Figure 10-2 shows the schematic diagram, and Figure 10-3 the stripboard layout.

The project is built on a small piece of stripboard, and mounted in a plastic project box. One hole in the project box allows light from the laser or flashlight to enter and fall on a diffuser, which illuminates the phototransistor. Another hole lets the sound from the loudspeaker escape.

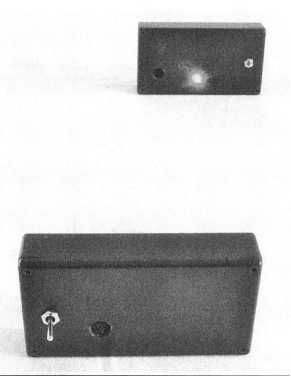

Figure 10-1   Sound over a laser beam

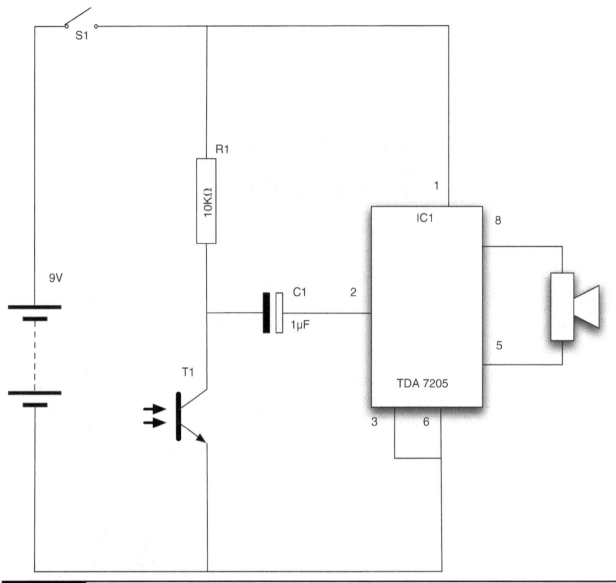

Figure 10-2 The receiver schematic diagram

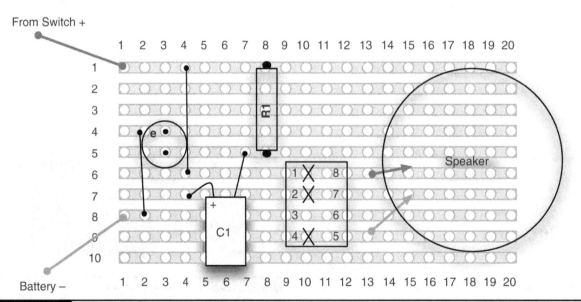

Figure 10-3 The receiver stripboard layout

What You Will Need

You will need the components in the Parts Bin to build the receiver.

You will also need the following tools to build the receiver.

T O O L B O X
■ Soldering equipment
■ A hot glue gun or self-adhesive pads
■ A drill and assorted drill bits

Assembly of the Receiver

We will assemble the receiver first, building the design onto a small piece of stripboard. The following steps will lead you through the process of assembling the receiver.

Step 1. The Stripboard

Cut the stripboard to produce an area of ten strips, each with 20 holes. You can do this with a strong pair of scissors or by scoring both sides of the board with a craft knife and then breaking it over the edge of a table. Use a drill bit twisted between your fingers to cut the gaps in the track.

Three cuts should be made in the track (Figure 10-4). Make these cuts using a drill bit as a hand tool, rotating it between your finger and thumb.

Step 2. Solder the Links

As always with stripboard, start by soldering those things closest to the board. In this case, that would be the two linking wires between tracks.

Figure 10-5 shows the stripboard with the link wires soldered into place.

Step 3. Solder the Resistor and IC

Next, solder the resistor.

Before soldering the IC, place the IC in the correct position, ensuring it is the right way around (the little dot indicating pin 1 should be top left). You may choose to use an IC socket to avoid damaging the IC by overheating it during

PARTS BIN			
Part	**Quantity**	**Description**	**Source**
T1	1	Phototransistor	Farnell: 1497673
R1	1	10kΩ 0.5-W metal film resistor	Farnell: 9339787
C1	1	1µF electrolytic capacitor	Farnell: 1236655
IC1	1	TDA7205 1-W amplifier	Farnell: 526198
S1	1	Toggle switch	Farnell: 1661841
Stripboard	1	Stripboard: ten strips, each with 20 holes	Farnell: 1201473
Battery clip	1	PP3 style battery clip	Farnell: 1183124
Battery	1	PP3 Battery	Hardware store
Socket	1	Optional eight-pin DIL IC socket	Farnell: 11011345
Loudspeaker	1	Miniature loudspeaker (8Ω)	Farnell: 1300022
Box	1	Plastic project box	Farnell: 301474
Diffuser		White expanded polystyrene packing material	
		Conductive foam IC packaging	

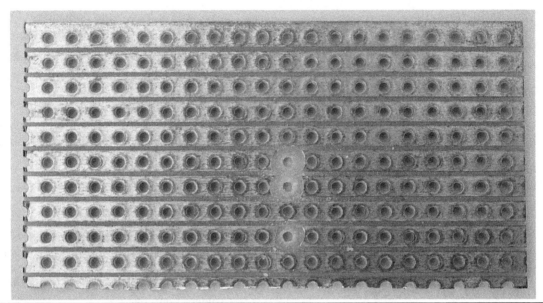

Figure 10-4 The stripboard for the receiver with tracks cut

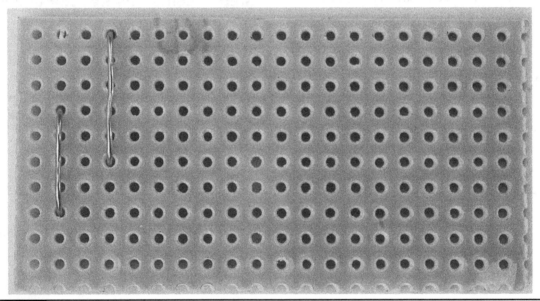

Figure 10-5 The receiver stripboard with the link wires in place

soldering. If you decide to solder the IC directly, try to minimize the time you are heating the pins. Also, if you have a difficult connection that takes a while for the solder to flow, wait until the IC is cool again before soldering the next lead.

Figure 10-6 shows the stripboard with the resistor and IC in place.

Step 4. Solder the Remaining Components

We can now solder in the rest of the components. Attach the capacitor so it lies on its side, reducing the height of the board.

When finished, our stripboard should look something like Figure 10-7.

Figure 10-6   The receiver stripboard with the resistor and IC

The loudspeaker has two wires soldered to its terminals. It is then glued onto the board and the leads connect through onto the stripboard.

Before soldering the phototransistor, make sure you have it the right way around. The longer lead is the emitter, which should be the higher lead on the board.

Step 5. Wire Everything Together

Using Figure 10-8 as a reference, wire up the project so we can test it before fitting it into a project box.

First, cut the red lead of the battery clip in half and strip and tin both ends of the cut lead. Then, solder both red leads to the switch and attach the

Figure 10-7   The completed stripboard for the receiver

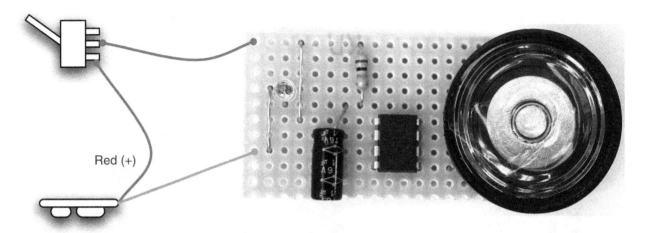

Figure 10-8   The receiver wiring diagram

loose end of the lead from the switch to the positive connection on the stripboard.

Solder the black negative lead from the battery clip to the negative connection on the stripboard. That's it. You're done. Your receiver should now look like Figure 10-9.

We cannot test the receiver properly without a transmitter, but if you connect a battery and switch the receiver on, you should hear crackling noises from the loudspeaker whenever you wave your hand between the phototransistor and a light source.

Step 6. Box the Receiver

If the receiver makes clicking noises when we flash a bright light into it, we can be fairly confident that it works okay, and go ahead and fit it into a project box.

It is a good idea to arrange the components in the box first, working out which places are best for them (Figure 10-10).

From these positions, work out where to drill the three holes we will make in the box lid: two large

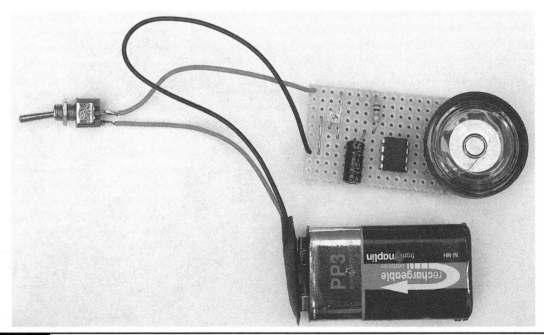

Figure 10-9   The receiver ready for testing

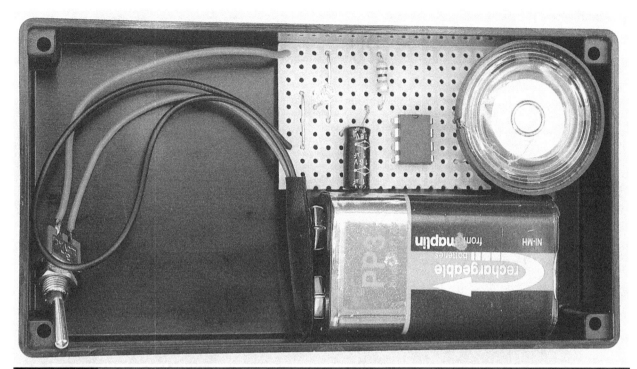

Figure 10-10   Laying out the parts of the receiver

holes for the sound from the speaker and the light sensor, and one small hole for the toggle switch.

Drill the holes in the lid and gently file off any burrs or rough edges in the plastic. The end result is shown in Figure 10-11.

Bare holes in the lid would look a little ugly, and also require a direct hit on the phototransistor by the laser, so let's fix some material on the inside of the lid behind each hole (Figure 10-12).

The diffuser is made from the same white expanded plastic packing material we used in Chapter 5. You can test its diffusing effect by touching a flashlight to one side of it. Looking from the other side, you will see that the material is all lit up, with the brightest area in the center. This effectively increases the target area that the laser has to hit from the tiny size of the phototransistor to the size of the hole in the lid.

The material covering the hole for the speaker was simply made from some of the black conductive foam that the IC was packaged in. You could use any thin material that will let the sound out unimpeded.

Figure 10-11   Figure 11: Drilling the box lid

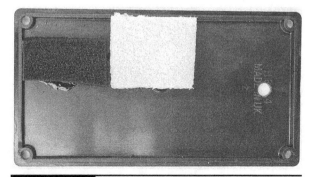

Figure 10-12   Covering up the holes

In both cases, the material is just glued in place. As an alternative, you could use a headphone socket in place of the speaker.

The stripboard should be fixed in place using glue or self-adhesive pads. The box that the author used was just the right width to hold the battery in place, so there was no need to secure it.

All that remains is to fit the switch, attach the battery, and screw down the lid of the box. The final boxed receiver is shown in Figure 10-13.

The Laser Transmitter

The laser transmitter will take the signal from a microphone and amplify it to modulate the power

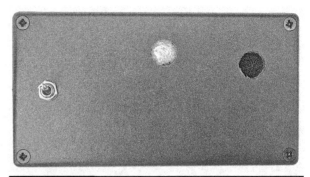

Figure 10-13 The completed receiver

going into a laser module. The schematic diagram for this is shown in Figure 10-14.

More information on how this works can be found in the "Theory" section at the end of this chapter. But the basic idea is that the signal from

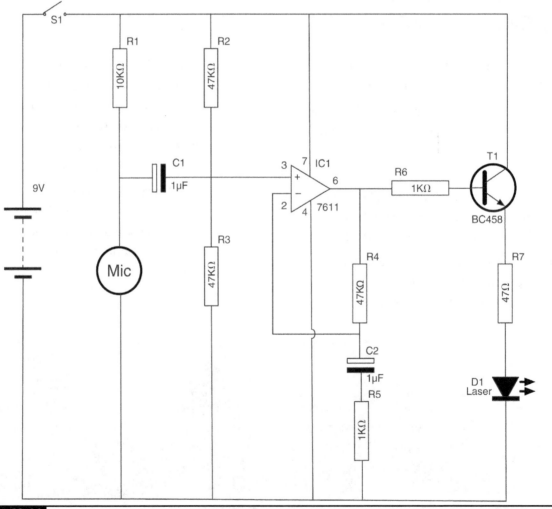

Figure 10-14 The schematic for the laser transmitter

the microphone makes the laser shine brighter or dimmer depending on the volume of the sound.

What You Will Need

You will need the following components to build the laser transmitter. They are listed in the Parts Bin.

You will also need the following tools to build the receiver.

TOOLBOX
■ Soldering equipment
■ A hot glue gun or self-adhesive pads
■ A drill and assorted drill bits

Assembly of the Transmitter

Figure 10-15 shows the stripboard layout for the transmitter. It is a little more complex than the receiver. It uses an operational amplifier IC and a transistor to increase the signal from the microphone to a level that is large enough to drive the laser.

The construction of the transmitter is described in the following step-by-step instructions.

Step 1. The Stripboard

The first step is to cut a piece of stripboard that has ten strips, each with 22 holes. There are then six breaks to be made in the tracks. Make these using a drill bit as a hand tool, rotating it between your finger and thumb. Figure 10-16 shows the prepared board, ready for the components to be soldered to it.

PARTS BIN			
Part	**Quantity**	**Description**	**Source**
R1	1	10kΩ 0.5-W metal film resistor	Farnell: 9339787
R2–R4	3	47kΩ 0.5-W metal film resistor	Farnell: 9340637
R5, R6	2	1kΩ 0.5-W metal film resistor	Farnell: 9339779
R7	1	47Ω 0.5-W metal film resistor	Farnell: 1127951
C1, C2	2	1μF electrolytic capacitor	Farnell: 1236655
T1	1	BC458 NPN transistor	Farnell: 1467872
IC1	1	7611 CMOS Operational Amplifier	Farnell: 1018166
D1	1	5 mW red laser diode module	eBay
IC socket		Optional: Eight-pin DIL IC socket	Farnell: 1101345
Mic	1	Electret microphone insert	Farnell: 1736563
S1	1	Toggle switch	Farnell: 1661841
Stripboard	1	Stripboard: 10 strips, each with 22 holes	Farnell: 1201473
Battery clip	1	PP3 style battery clip	Farnell: 1183124
Battery	1	9V PP3 battery	
Box	1	Plastic project box	Farnell: 301474

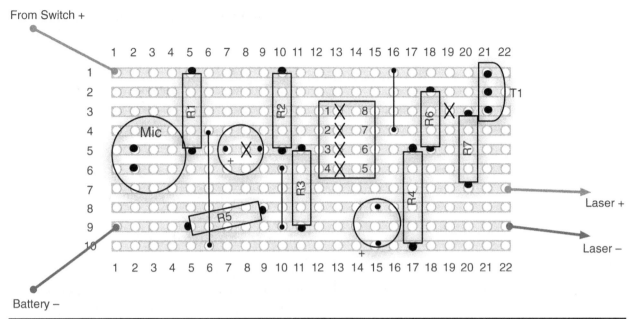

Figure 10-15 The stripboard layout for the transmitter

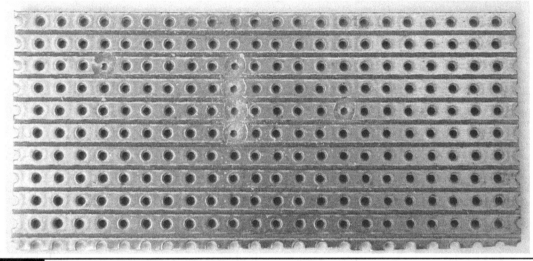

Figure 10-16 The stripboard for the transmitter

Step 2. Solder the Links

Before starting on the real components, solder the three wire links into place. When this is complete, your board should look like Figure 10-17.

Step 3. Solder the Resistors

We can now solder all the resistors in place. Take special care with R5, which is soldered at an angle across one of the links. The leads of the resistor

should be kept away from the wire of the link. However, it does not matter if the body of the resistor touches the link.

When all the resistors are soldered into place, your board will look like Figure 10-18.

Step 4. Solder the IC

Place the IC in position (Figure 10-19), ensuring it is the right way around—the little dot indicating pin 1 should be top left. Solder the IC into place,

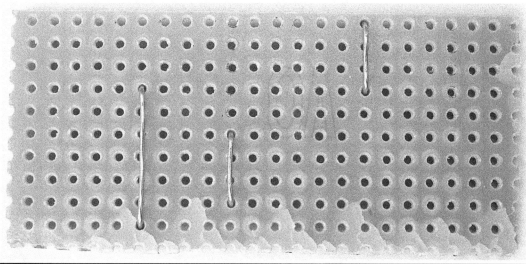

Figure 10-17 The transmitter stripboard with links

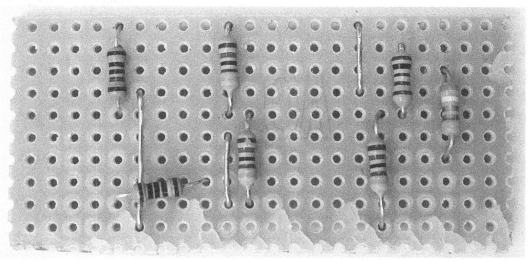

Figure 10-18 The transmitter board with resistors

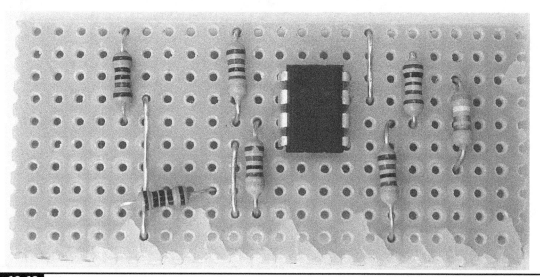

Figure 10-19 The transmitter board with IC

using the same precautions as with the receiver's construction. Just like with the receiver, you may decide to use an IC socket rather than solder the IC directly.

Step 5. Solder the Remaining Components

The rest of the components can now be soldered onto the board. Solder the transistor first (making sure it is the right way around), and then the capacitors and the electret microphone insert.

The microphone that the author used was designed to be soldered to a PCB. However, you may have to solder short leads to the contacts on the microphone.

The microphone must also be connected the correct way around. The contact that is also connected to the case should be the negative lead, which should be the lower contact on the stripboard. You may need to test this with a multimeter. If the transmitter does not vary the brightness of the laser, you may need to try the microphone insert the other way around.

Figure 10-20 shows the completed stripboard.

Step 6. Wire Everything Together

Using Figure 10-21 as a reference, wire up the project so we can test it before fitting it into a project box.

First, cut the red lead of the battery clip in half and strip and tin both ends of the cut lead. Then, solder both red leads to the switch and attach the loose end of the lead from the switch to the positive connection on the stripboard.

Next, solder the black negative lead from the battery clip to the negative connection on the stripboard.

Shorten the leads on the laser module so they are about 2 inches (50mm) and attach them to the stripboard, ensuring the polarity is correct.

Step 7. Test the Transmitter

Attach the battery and aim the laser onto some white paper. Note that if you tap the microphone or speak loudly into it, the brightness will alter. If this does not happen, then check your circuit and make sure the microphone is the right polarity.

Figure 10-20 The completed transmitter stripboard

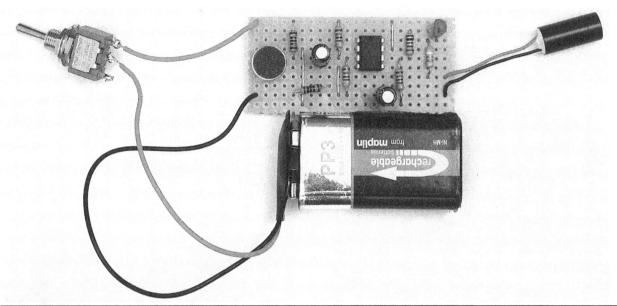

Figure 10-21 Wiring the transmitter

If you now turn on the receiver, you will probably find that when you shine the laser onto the diffuser you get a bit of feedback howl. If this happens, it's a very good sign and you can try out the project properly by taking the transmitter away from the receiver and setting it to point at the diffuser. This normally requires careful alignment. The author found it useful to prop things up with stacks of coins, taken from the pockets of his minions (the Evil Genius does not carry cash).

Step 8. Box the Transmitter

Now that we are confident that everything works, we can box up the transmitter in much the same way we did the receiver.

Start by laying out the components in the box (Figure 10-22). In this case, the laser module will stand vertically against one of the plastic pillars to which the lid screws attach. A hole will be drilled in the bottom of the box for the light to emerge.

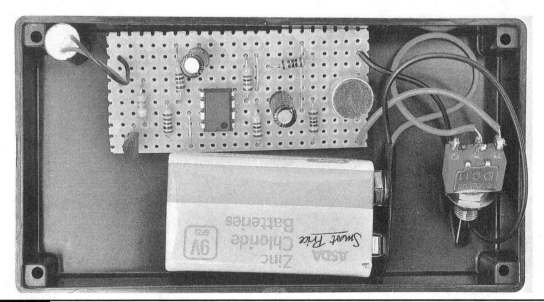

Figure 10-22 Laying out the transmitter

Two other holes are required: one for the microphone sound to enter through, and one for the on/off switch. Figure 10-23 shows the approximate position and sizes of the holes.

Figure 10-23 Drilling the transmitter box

Stick some thin foam or other material behind the microphone hole, just as you did for the speaker hole in the receiver (Figure 10-24).

Make sure the laser module lines up with the hole beneath it and glue it into place against the pillar. Self-adhesive tabs or a blob of glue will fix the stripboard into place (Figure 10-25).

Figure 10-26 shows the transmitter, fully assembled.

Figure 10-24 Foam material behind the microphone hole

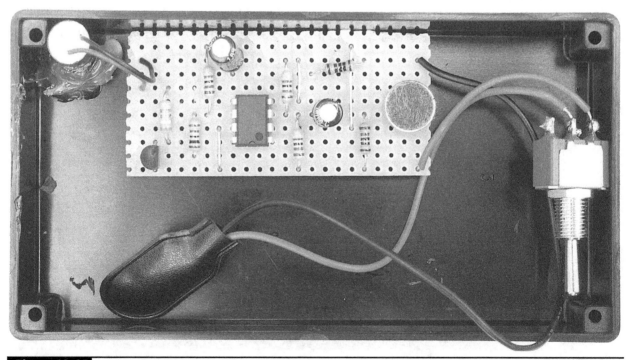

Figure 10-25 Fixing everything in place

Figure 10-26 The fully assembled transmitter

Using the Project

Alignment

It is surprisingly difficult to get a laser dot to hit the diffuser on the receiver from a reasonable distance. Ideally, both the receiver and transmitter should be fixed in place. One way to do this is to rest the receiver on the window sill of the Evil Genius' friend and mount the transmitter on a tripod. You can do this by drilling an additional hole in the transmitter case that is slightly smaller than the bolt on a tripod. You can then carefully screw the tripod into the case, which will "self-tap" a thread into the hole.

Be careful not to tighten the bolt too tight. For a more solid solution, you could drill a bigger hole and glue a nut to the inside of the transmitter.

Actually, aligning the transmitter with the receiver is just a case of careful aiming from the transmitter.

Time Travel

One very spooky effect of this project is that if you have the whole apparatus in one room and someone standing at the transmitter claps their hands, the person at the receiver will hear the clap come through the loudspeaker of the receiver *before* they hear the real clap from the person!

A moment's thought will tell you that this is just because light travels faster than sound, but it is very strange to observe, because it does seem like the receiver is anticipating the clap.

Duplex

As we mentioned in the introduction, the Evil Genius is not overly interested in anything that anyone else has to say; however, if you wish to have two-way communication, you will need to make two receiver/transmitter pairs.

When setting them up, you will have to avoid the problem of feedback. Feedback occurs when the sound entering the first transmitter is received in the first receiver, whose speaker is audible to the microphone of the second transmitter, whose output then goes back to the first receiver, and around and around forever.

It's a bit like the audio equivalent of standing between two mirrors that are opposite each other. While it does produce some interesting sounds, it will not help you hear what people are saying.

You can avoid this feedback by keeping the receiver and transmitter a little way apart.

Theory

This is a good project for learning a bit more about electronics. It covers amplification and modulation, two components needed for almost every type of electronic communication.

Amplitude Modulation

If you have a radio, it will probably have a switch somewhere that lets you select AM or FM. These acronyms stand for Amplitude Modulation and Frequency Modulation. Although AM and FM operate on different frequencies in public broadcasting, it is not the choice of frequency that differentiates AM from FM, but the way in which the sound signal is sent over the airwaves.

Listening to AM and FM channels, you have likely noticed that AM is of much lower quality than FM, and is generally used for voice rather than music for this reason.

This project uses AM to send the sound over the laser. Figure 10-27 shows the oscilloscope trace for both the voltage across the laser diode (top trace) and the signal from the microphone (bottom trace).

Looking at the voltage across the laser diode, we can see that the channel has a sensitivity of 1V per square on the screen. So, the voltage across the laser varies from 2.2V to nearly 4V. This is quite an exaggerated swing, because the sound source was loud and placed close to the microphone. Under normal usage, the signal will be a little smaller and also suffer less from the obvious distortion of the shape of the waveform.

Our input signal is "modulating" the voltage across the LED, and hence its brightness in time with the signal, which in this case is a frequency of 2 kHz.

When the beam of the laser arrives at the phototransistor, that same variation in the brightness will be amplified and used to drive a loudspeaker, re-creating the sound.

When AM is used in a radio, the approach is slightly different in that there is a carrier wave at a

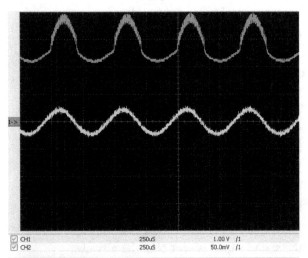

Figure 10-27 Oscilloscope trace for AM on the laser

certain frequency (the frequency you set your dial to), and it is this wave whose amplitude or strength is modulated.

AM suffers from the fact that the signal is carried by variations in its strength. This makes it very susceptible to any other factors that may change the amplitude. In radio, that might be changes in atmospheric conditions or obstacles getting in the way of the signal.

The far superior FM, or frequency modulation, does not modulate the strength of the signal. Instead, it alters the frequency of the carrier signal a little. You can find excellent resources on the Internet explaining frequency modulation.

How the Transmitter Works

The signal coming from the microphone is far too small to directly alter the power of the laser diode. To amplify the signal, we use a type of amplifier IC called an operational amplifier.

An operational amplifier will not, on its own, provide a high-output current. It is solely concerned with amplifying the voltage or amplitude of the signal.

An operational amplifier can be used in many different ways, and the configuration used in this project is as a non-inverting amplifier. The amplifier has two inputs, a positive input and a negative input. It is the difference between these two inputs that is amplified. On its own, the amplification or "gain" of an operational amplifier is very large indeed, and may be as much as a million—thus, a signal of just 1μV would be amplified to become 1V. In practice, this is far too much, so the gain is reduced using feedback.

The schematic diagram for the project is repeated here in Figure 10-28 for convenience.

The signal from the microphone is connected to the positive amplifier input. This input is basically held halfway between GND (0V) and the supply voltage of 9V by R2 and R3; however, it will

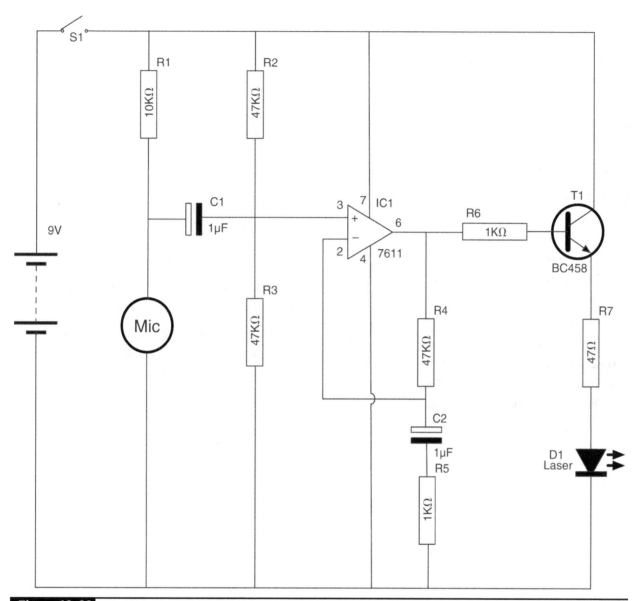

Figure 10-28 The schematic diagram for the transmitter

fluctuate from this by the influence of the microphone.

A portion of the output determined by R4 and R5 is fed back into the negative input of the amplifier to reduce the gain. The formula for calculating the gain is:

$$gain = (R4 + R5) / R5$$

which in this case will be:

$$(47 + 1) / 1 = 48$$

The signal at the input will therefore be multiplied by a factor of 48.

So, we have solved the problem of increasing the amplitude or voltage of the signal, but not the problem of providing sufficient current to drive the laser diode. To do this, we use a transistor in a mode called emitter-follower. That is, the voltage at the emitter (indicated by an arrow on the transistor circuit symbol) tracks or follows the voltage at the base (actually, 0.5V less). It will actually do this no matter how much current flows through the collector emitter of the transistor, providing current amplification rather than voltage amplification.

R7 is there to limit the current to the laser diode.

How the Receiver Works

Most of the work in the receiver is carried out using a single audio amplifier IC. This device, the TDA 7205, will amplify the signal from the phototransistor. The capacitor C1 provides AC coupling—that is, it only allows the changes to the microphone to pass through to the amplifier, not any DC bias.

The phototransistor allows more or less current to flow depending on the light falling on it. The more light, the more it conducts. This current will flow through both the phototransistor and the resistor R1. It is the voltage dropped across R1 that will be amplified by the IC. So, using a higher value of resistor here would increase the sensitivity of the receiver; a lower value would decrease the sensitivity. The value of 10kΩ was found to be about right.

Summary

This has been an interesting project, in which we have learned something about amplification and modulation—the foundations of radio communication. While the end result is not as useful as a walkie-talkie or a cell phone, it does work, at least over a fairly short range.

In the next chapter, we can look forward to a project that the Evil Genius likes to use to torment his minions: flashing a camera flash in their faces.

CHAPTER 11

Flash Bomb

PROJECT SIZE:	Small
SKILL LEVEL:	★★☆☆

TIME SOMETIMES PASSES SLOWLY in the Evil Genius' Lair. The winter months drag on with little opportunity for evil. The long dark evenings can be enlivened by inflicting pranks on the hapless minions.

The Evil Genius likes to amuse himself by leaving an interesting looking object lying around. When picked up by a passing minion, suspecting it might be something that has been misplaced by its master, it administers a bright flash into their faces.

The device uses a secret micro-switch on its base to detect when it has been picked up.

This is an easy project to make as it is built using the flash module of a single-use camera, which is then housed in a small plastic food container (Figure 11-1).

WARNING!

This project uses a bright xenon flash tube. Because of this, you should take the following precautions.

- When disassembling the disposable camera, be extremely careful since the flash capacitor is charged to over 300V and the trigger transformer produces voltage spikes of 4kV.

- Always disconnect the battery and discharge the flash before touching the circuit board.

Figure 11-1　The flash bomb

- Do not use this project around anyone who suffers from epilepsy.

- Do not stare into the flash tube—it is very bright at close range.

What You Will Need

You will need the following components to build this project. They are listed in the Parts Bin on the next page.

The author used a Kodak single-use camera. If you are using a different camera, the design will differ from the following instructions, but you should still be able to adapt the design.

PARTS BIN			
Part	**Quantity**	**Description**	**Source**
	1	Single-use camera with flash (Kodak)	
S1	1	Miniature micro-switch	Farnell: 1735350
	1	Small plastic food container	

With the advent of digital cameras, it is becoming harder and harder to find single-use cameras. You can often obtain used camera bodies for free from film processors. If you cannot, a new one will only cost you a few dollars.

You will also need the following tools for this project:

TOOLBOX
■ Soldering equipment
■ A hot glue gun or self-adhesive pads
■ An electric drill and assorted drill bits

Assembly

We will remove the main circuit board of the camera and mount it in a transparent food container. The contacts normally triggered by the camera's shutter will be replaced by a micro-switch. We use the normally closed contacts of the switch so that when the project is sitting on a table, the switch will be depressed; however, as soon as it is picked up, the contact is made and the flash fires.

The schematic diagram for the project is shown in Figure 11-2.

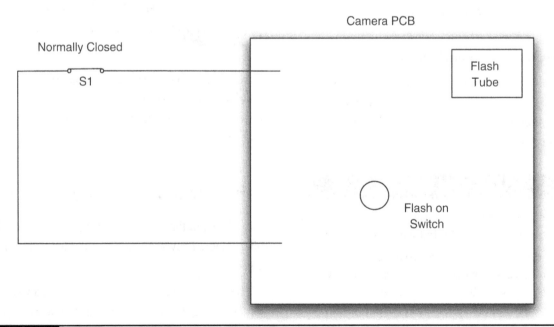

Figure 11-2 The schematic diagram for the project

Step 1. Disassemble the Disposable Camera

These disposable cameras (Figure 11-3) are remarkably complex considering their price tag of a few dollars. Little difference exists between them and a cheap film camera. In the device chosen, the flash uses a decent-sized capacitor and packs a considerable punch.

Do not take the camera apart if the flash is on. The camera will have a switch that turns it on, and when the flash is charged, a ready light will come on. The camera used by the author had a press switch that had to be held down for eight seconds to turn the flash on. If this ready light is on, its capacitor will be charged up to over 300V, just waiting to discharge itself through you. This will hurt and could cause some injury. Not only that, but many parts of the circuit board will be at high voltage even though the flash might be switched off.

So, make sure the flash is off. If you have had the flash on to try out the camera, turn it off and leave the camera alone for a few hours. Even then,

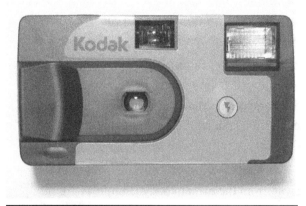

Figure 11-3   A single-use camera

use caution with the following steps until you have had a chance to discharge the capacitor.

If you peel back the adhesive labels over the plastic, you will find clips that hold the camera in place. Carefully unclip these, revealing the innards of the camera (Figure 11-4).

Notice the large tubular capacitor below the lens. Before you go any further, without touching anything metal, flip out the single AA battery, and using a screwdriver, short out the two leads of the

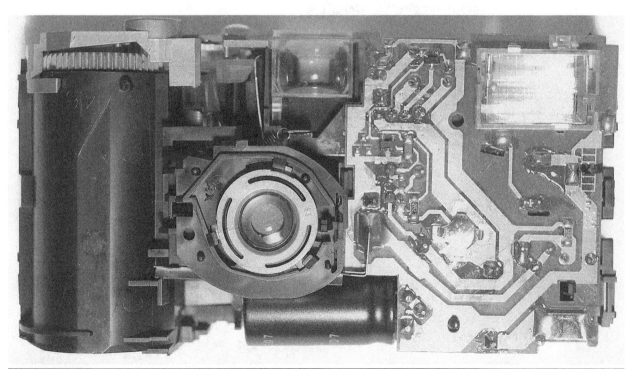

Figure 11-4   The inside of the single-use camera

capacitor. If the capacitor was charged, there will be a big bang and sparks, and your screwdriver will have scorch marks on it.

However, if you have followed the instructions and only disassembled the camera when discharged, you should get no more than a little pop as the residual charge in the capacitor is discharged.

We can now carefully pull the circuit board away from the camera body (Figure 11-5).

Looking at Figure 11-5, we can see the back of the flash tube on the bottom right. Be careful not to touch the glass of the tube because it will shorten its life. On top of the flash tube is one end of the battery clip; the other end of it is shown in the top right of Figure 11-5. Note that it is not immediately obvious which is the positive connection for the battery. Actually, the top connector is the positive connector.

If you are using a different type of camera, then it is worth taking note of the battery polarity when disassembling the camera   .

Directly beneath the capacitor you should see a long metal contact. Directly beneath that is a second shorter contact. It is these contacts that are closed by the shutter when it opens to allow light to reach the film.

Step 2. Attach the Micro-switch

We can now cut off those large metal contacts and replace them with a micro-switch attached to a short length of twin wire. Two separate wires are fine if you do not have twin core flex available.

We need to attach our micro-switch so that the contacts are "normally closed." That is, the connections to the switch are normally closed, but when you press the lever at the top of the switch, they open.

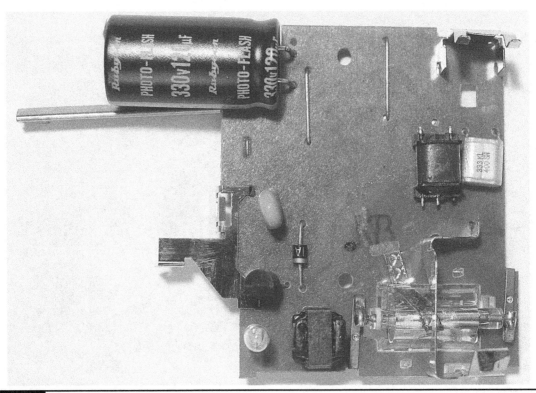

Figure 11-5   The camera's PCB

If you look closely at the micro-switch, its three pins will normally be labeled Common, NC (normally closed), and NO (normally open). Solder your leads to the NC and Common connections. If your micro-switch is not labeled in this way, then test it with the continuity setting on your multimeter to determine the normally closed and common connections.

When it's all soldered together, you should have something that looks like Figure 11-6.

In the center of the circuit board, you can see a little raised area with four terminals. This is the switch that has to be held down for eight seconds to turn on the flash. You will find that when you push it with a screwdriver, it will click.

At this point, it's a good idea to test what we have done so far, before we start fixing it into its box.

Again, exercise extreme caution, or you will get a nasty jolt. In other words, don't touch anything metal with your fingers.

Carefully fit the battery back in. Nothing will happen immediately since the flash has not been turned on. Before you do that, hold down the lever on the micro-switch and then with the other hand (using something insulating such as plastic) press in the switch on the circuit board for eight seconds.

You should hear the charging circuit kick into action and after a few seconds the indicator light will glow.

At this point, if you release the micro-switch, you will get a flash.

Before moving onto the next step, we need to make sure the capacitor is discharged. The best way to do this is to, without touching the circuit board, take the battery out, and then discharge the flash by pressing and releasing the micro-switch again so the capacitor is discharged through the flash tube.

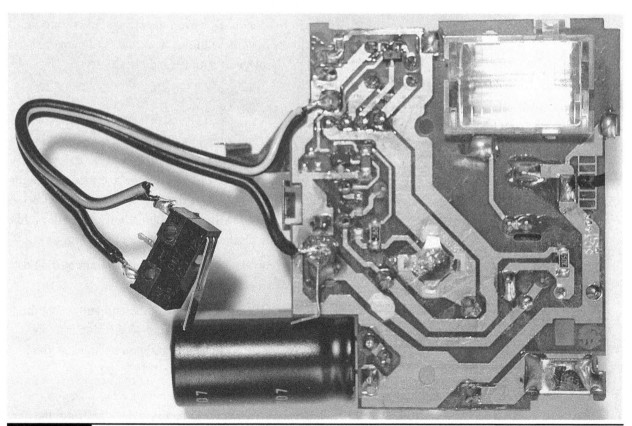

Figure 11-6   The micro-switch attached to the PCB

Step 3. Box the Project

As you can see from Figure 11-1 earlier, the PCB is simply glued to the bottom of the plastic food container using a hot glue gun. But before doing that, we need to drill a small hole opposite the push switch that turns the flash on. We will then be able to poke a toothpick in it to turn the device on.

The glue will stick better if the plastic container is roughened with sandpaper.

Figure 11-7 shows the open box with the PCB glued inside.

The micro-switch is fitted into the lid of the box by cutting a slit in the lid, pushing the switch through, and then fixing it in place with the hot glue gun (Figure 11-8).

We can now test the project by putting a toothpick through the hole to turn the flash on while holding down the micro-switch. When the indicator light comes on, we release the switch and it should flash.

Using the Project

For best results, leave the project lying around somewhere slightly out of place, where someone is likely to pick it up and look at it.

Figure 11-8   The micro-switch attached to the box lid

Theory

Modding the single-use camera in this way has saved us the trouble of designing the electronics from scratch. However, it's always interesting to know how these things work, and this camera is a masterpiece of simple (and cheap) design.

Flash Guns

A flash gun works much like the coil gun we built in Chapter 1. In both cases, we charge up capacitors over a period of time and then discharge them in a very short time. In the case of the coil gun, we discharge it through a hefty coil, but in this case we discharge it through a xenon flash tube.

Figure 11-9 shows a logical diagram of the flash circuit.

Another difference with the coil gun is that the coil gun operated at around 40V (using four 9V batteries), whereas the flashgun operates at about 330V, using just a single 1.5V battery.

This is only possible because the flashgun contains a circuit to increase the 1.5V from the

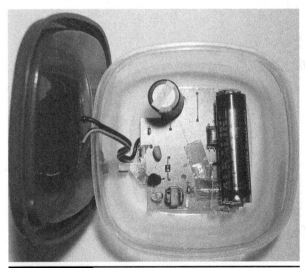

Figure 11-7   The PCB fitted inside the box

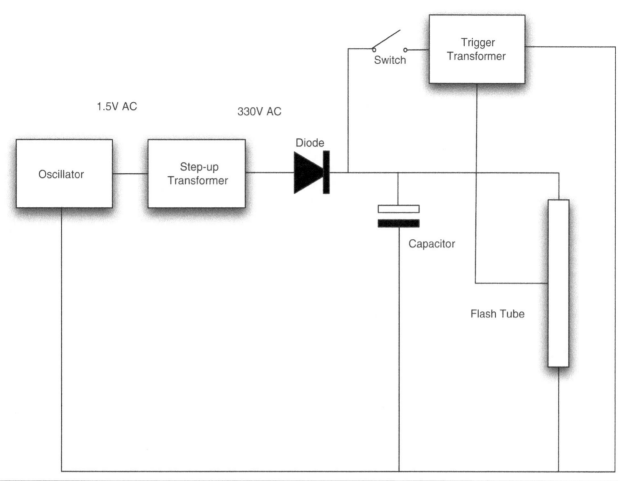

1.5V AC

330V AC

Oscillator

Step-up Transformer

Diode

Switch

Trigger Transformer

Capacitor

Flash Tube

Figure 11-9   Logical diagram for the flash gun

battery to 330V. It does this using an oscillator that drives a transformer.

The high voltage is then rectified by a diode and charges the capacitor. Even when the capacitor is fully charged to 330V, the flash tube will not fire until a much higher trigger voltage pulse of around 4kV is applied to the trigger connection of the tube. This pulse is generated by a special high-voltage trigger transformer. The trigger transformer receives a short pulse when the shutter is released.

Summary

This is a nice easy project and one intended for mischief. However, please be careful, both of the high-voltage electronics and in using it responsibly.

In the next chapter, we stay with the flashing theme and develop a project that uses 36 high-brightness LEDs to make a strobe light.

CHAPTER 12

High-Brightness LED Strobe

PROJECT SIZE:	Medium
SKILL LEVEL:	★★★★

THE EVIL GENIUS FINDS that a bright strobe light can be used to strike fear into the heart of his enemies. (See Figure 12-1.) What's more, it's great for parties. The Evil Genius writes himself a memo: "Arrange party for me. Order minions to capture some guests."

This project uses an array of 36 high-brightness LEDs (HB LEDs). It's small, battery-powered, and very efficient. A knob on the side allows the frequency of the flashing to be varied.

WARNING!

Because this project creates a bright strobe light, you should be aware of the following dangers.

- Do not use this strobe light around anyone who suffers from epilepsy.
- Do not stare into the strobe light—it is very bright at close range.

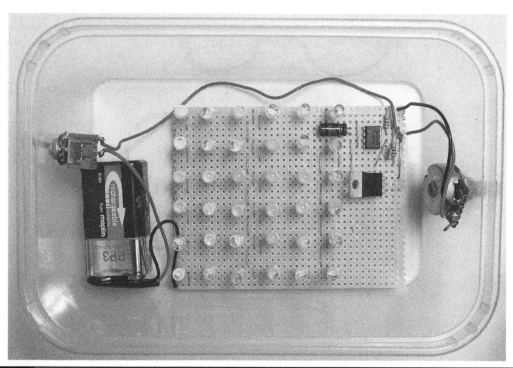

Figure 12-1　High-power strobe

What You Will Need

You will need the following components to build this project. They are listed in the Parts Bin.

When buying the LEDs, look first for very high-brightness white LEDs. Those of around 20,000 mcd will generally have a forward voltage of around 3.3V, but check this on the description of the product before you order. If the forward voltage is not given in the details, look at another listing—there are usually dozens. Buy a single lot (often sold in 50s) since you will be more likely to get near-identical LEDs that way. In any case, buy some spares, because I have occasionally found some duds in these cheap bulk offerings.

You will also need the following tools for this project that are listed in the Toolbox.

TOOLBOX
■ An electric drill and assorted drill bits
■ Two pencils
■ Soldering equipment
■ A hot glue gun or self-adhesive pads

Assembly

Figure 12-2 shows the schematic diagram for the project. The heart of the project is a 555 timer IC. This generates pulses that turn on a MOSFET transistor that drives the LEDs.

We first met the 555 timer IC in Chapter 4, where we used it to generate pulses for our servo laser turret. If you are interested in how this works, look at the "Theory" section at the end of Chapter 4.

PARTS BIN			
Part	**Quantity**	**Description**	**Source**
R1	1	1MΩ variable resistor	Farnell: 1629581
R2, R3	2	10kΩ 0.5-W metal film resistor	Farnell: 9339787
C1	1	1μF 16V	Farnell: 1236655
C2	1	10nF	Farnell: 1694335
T1	1	P-Channel MOSFET FQP27P06	Farnell: 9846530
IC1	1	NE555 timer	Farnell: 1467742
D1-36	36	White LED 5mm 20,000 mcd, I_f 30mA V_f 3–3.3V	eBay
IC socket	1	Eight-pin IC socket (optional)	Farnell: 1101345
S1	1	SPST on-off toggle switch	Farnell: 1661841
Box	1	Food container with transparent lid	Supermarket
Knob	1		Farnell: 1441137
Battery clip	1	PP3 battery clip	Farnell: 1650667
Battery	1	9V PP3 battery	

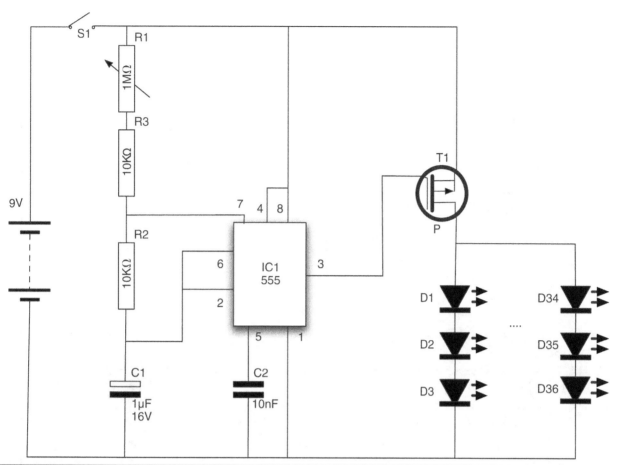

Figure 12-2   The schematic diagram

The LEDs are arranged in a six-by-six pattern on the main area of the stripboard, with an area of the board reserved for the transistor 555 timer and other components.

The following step-by-step instructions lead you through making the strobe light. Everything apart from the switch, battery clip, and variable resistor are built onto a piece of stripboard.

Step 1. Prepare the Stripboard

Cut the stripboard to give an area of 29 strips, each with 38 holes. You can do this with a strong pair of scissors or by scoring both sides of the board with a craft knife and then breaking it over the edge of a table.

The stripboard tracks will need breaking at the places marked with an X on Figure 12-3. Note that Figure 12-3 shows the board from the top. To translate the positions to where the track is to be cut underneath, it can be useful to push a piece of wire through from the top.

Many breaks need to be made, so check carefully to be sure you have made all of them.

The back of the stripboard with all the breaks made is shown in Figure 12-4.

Check the stripboard very carefully to make sure all the track breaks are in the correct place before you start soldering.

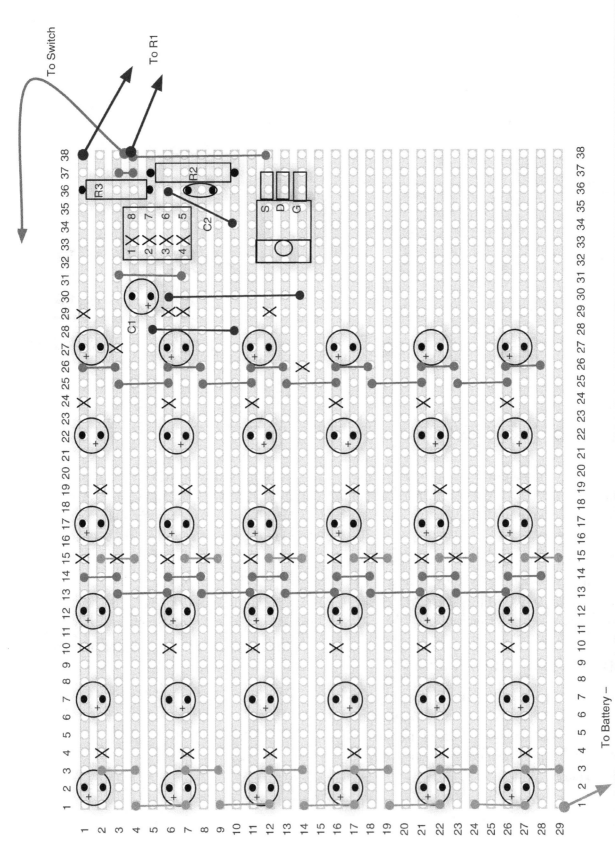

Figure 12-3 Stripboard layout

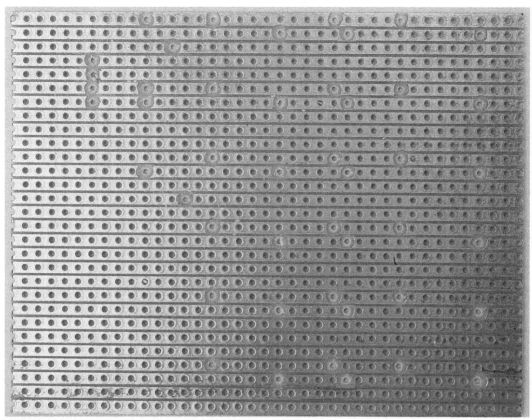

Figure 12-4   The stripboard with breaks

Step 2. Solder the Links

It is always best to solder the components that lie closest to the board first, so if you turn the board over and lay it on a flat surface while soldering, the component you are soldering will stay in place.

Start by soldering the links in place. Either use solid core wire with the insulation stripped off, or for the shorter links, you may find that the snipped-off leads from resistors soldered on previous projects are about the right length. When all the links are in place, your board should look something like Figure 12-5.

Step 3. Solder the Resistors and IC

Next, solder the resistors into place and then the IC or the IC socket. The advantage of using an IC socket is that it is far more tolerant of overheating during soldering than the IC itself. But if you are experienced at soldering and are confident you can solder in the IC with minimal heating, you may not wish to use the socket.

Be careful to get the IC the right way around (pin 1 with a dot next to it is at the top left of the IC).

Step 4. Solder the Transistor and C2

Bend the legs of the transistor after the point where the legs get thin. This will allow the transistor to lie at an angle to the stripboard (see Figure 12-6) so it does not stick up above the height of the LEDs. The transistor is designed to be bolted to a metal heat-sink, but in this project it will not consume enough power for this to be necessary.

We are not going to solder C1, the larger electrolytic, until we have soldered the other components into place, especially all the LEDs. However, we can attach C2 since it is small.

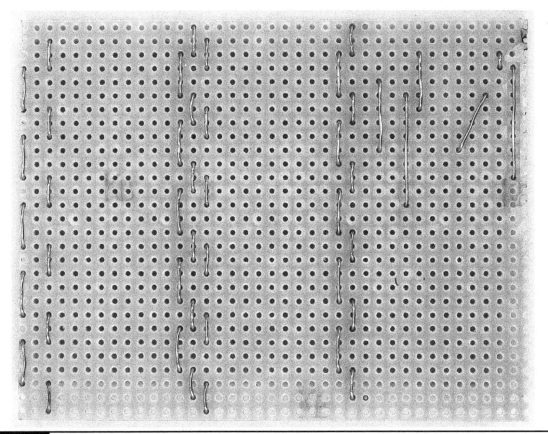

Figure 12-5   The stripboard with the links in place

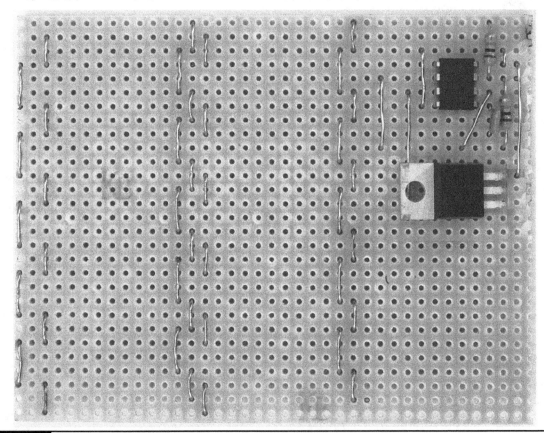

Figure 12-6   The stripboard ready for soldering the LEDs

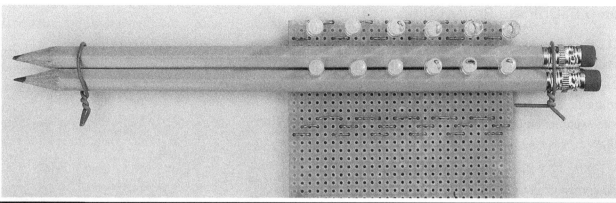

Figure 12-7   Soldering the LEDs

Step 5. Solder the LEDs

To make a neat job of it and ensure all the LEDs stand out from the board by about the same amount, use a spacer to prop the LEDs in the right place. Unfortunately, we cannot just put the LEDs all the way through the board, as this would mean we would be soldering very close to the body of the LED itself and could possibly damage the LEDs.

We will use two pencils as the spacers and solder a column of LEDs at a time. Figure 12-7 shows the second row of LEDs being soldered.

Line the pencils up next to each other and twist wire around each end. Make sure the wire at the sharp end of the pencil can be easily slipped off.

Put a row of LEDs loosely into the board and then push the board onto your work surface to lift the LEDs high enough off the surface of the board to insert the pencils on either side of the leads. Make sure you have the LEDs the correct way around; they should all be facing in the same direction for a particular column. However, beware, because each column alternates having the positive leads of its LEDs at the top. Use Figure 12-3 as a guide. The longer lead of the LED is always the positive lead.

Slip the wire back over the sharp end of the pencils and then turn everything over. Straighten up the LED leads before soldering the column of

LEDs into place. When you turn the board over, you may find an LED or two are not properly lined up. You can adjust the LEDs, bending them left or right, but do not try and bend then up and down, as you are likely to lift the copper track off the stripboard. Instead, hold on to the plastic lens of the LED, and at the same time melt the solder on one of the leads and adjust the position of the LED.

Repeat this procedure for all six columns of LEDs.

Finally, you can solder C1 into place, making your completed board look something like Figure 12-8.

Figure 12-8   The completed board

Step 6. Drill the Project Box

The project is contained in a plastic food container. The container has holes drilled for the switch and the variable resistor on the side. Since the LEDs will generate some heat, it is probably wise to drill some ventilation holes in the back of the box to allow excess heat to escape.

Figure 12-9 shows the box drilled and ready for the components to be fitted.

Step 7. Wire Up

Figure 12-10 shows how the stripboard variable resistor, switch, and battery clip are wired together. Wire them together outside of the box so you can test it before fitting everything into place.

Connect up the battery and make sure everything works. If some of the LEDs are not lit, turn it off and check the stripboard for any solder bridges between tracks. Also, check that you have the LEDs the right way around.

Step 8. Fit the Parts into the Box

The final step is to fit all the components into the box. To prevent the board and battery from moving

Figure 12-9   Drilling the box

around in the box, attach the stripboard to the box using a few blobs from a hot glue gun, or use some double-sided sticky pads. The battery is trickier, because you will need to be able to replace it. So, self-adhesive putty should do the trick.

Since the project has a moderately high current consumption (up to 150mA), it is probably worth using a rechargeable PP3 battery, or for a longer battery life, a battery holder pack that takes six AA or AAA batteries that is terminated in a PP3-type plug. However, be aware that this project will not work at voltages lower than 9V, and voltages higher than about 10V are likely to shorten the life of the LEDs.

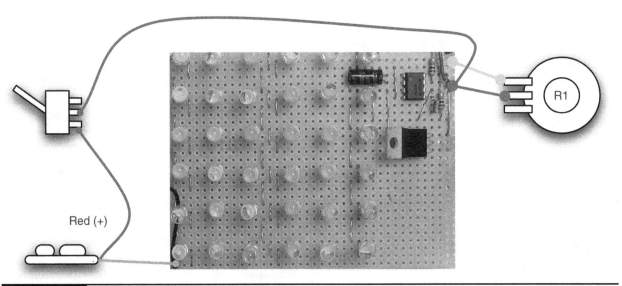

Red (+)

Figure 12-10   Wiring diagram

Theory

LEDs are the component of choice for generating light. In this section we take a closer look at LEDs and how to use them.

Using LEDs

LEDs have been used as indicator lights for many years. They have a great advantage over incandescent bulbs in that they last almost indefinitely (if treated well) and generate light efficiently without large amounts of waste heat.

In recent years, LED technology has improved so that all colors of light can be generated, and at intensities that make LEDs suitable for illumination rather than just indication. Indeed, where the author lives, LEDs are now being used in street lighting.

Although LEDs are pretty wonderful, they are more difficult to use than conventional incandescent bulbs. For a start, they have a polarity, so you have to drive them from DC and you have to connect them the right way around. Second, they cannot normally be connected directly to a power source like a battery. They require a resistor to limit the current; otherwise, the LED will draw too much current and burn out.

This design is typical of LED applications in that it requires a number of LEDs to achieve the desired brightness. In this case, quite a big number. So we have to work out how to wire up our LEDs so they shine brightly but without burning out. The key to this is to control the current that flows through the LED. This figure will typically be between 10mA to 30mA for a small regular LED and considerably higher for special-purpose LEDs such as the Luxeon LEDs, which can be up to a few watts in power. This figure goes under the name of I_f, or forward current. The other value we need to know regarding the LED is called V_f, or forward voltage. This is the voltage across the LED when it has I_f flowing through it.

The LEDs used in this project have a recommended I_f of 30mA, at which there will be a V_f of 3.3V. So to get them as bright as possible without damaging them, we need to set the current through them to be about 30mA. The problem is that batteries and power supplies are voltage sources rather than current sources. That is, you choose a battery that will maintain a more or less constant voltage of, say, 9V, no matter how much current you draw from it. But what we really want for LEDs is something that will maintain a current of 30mA even if the voltage supply goes up or down.

Some circuits can provide a current source in this way—many based on the L200 IC. The cheaper alternative is to use a current limiting resistor, as shown in Figure 12-11.

Figure 12-11 shows how if we know V (the supply voltage), V_f (the forward voltage of the

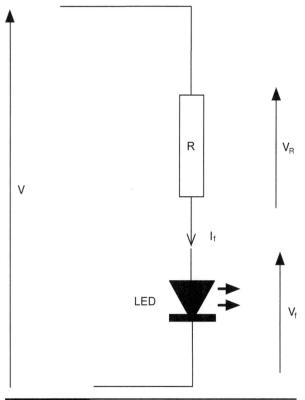

Figure 12-11 LED and current limiting resistor

LED), and I_f (the forward current of the LED), we can calculate a suitable value of resistor to use.

Since $V = V_f + V_R$, we know that $V_R = V - V_f$ and since the current through the resistor will be the same as the current through the LED (I_f), then using Ohm's law we have $R = (V - V_f) / I_f$.

For example, if we have a 9V battery and we want to drive an LED with a forward voltage of 2V and a forward current of 20mA, we need a resistor of value:

$$(9 - 2) / 0.02 = 350\Omega$$

So, you may be wondering how we are managing to drive all 36 of our LEDs without a single current-limiting resistor. The answer is that we are relying on the fact that if we drive three LEDs in series from 9V, each will drop 3V, which should still allow our LEDs to draw enough current to shine brightly without burning out. However, a 9V battery will often supply up to 10V

when it is new. This would give 3.3V per LED, which is still fine.

We then repeat this pattern six times to light all of our LEDs.

Summary

This is a great project if you like soldering. With 36 LEDs to solder, there are a lot of connections to be made.

As a way of driving LEDs, this design is far from ideal, and if the reader is interested in such things, they are encouraged to look at more advanced techniques for driving LEDs, such as constant current power supplies.

In the next chapter, we will turn our attention to gravity, and more specifically, how to defy it—with a levitation machine.

CHAPTER 13

Levitation Machine

PROJECT SIZE: Medium

SKILL LEVEL: ★★★☆

WHEN PLANNING CAMPAIGNS for world domination, the Evil Genius likes to have his globe of the Earth float in the air above his desk, suspended by invisible anti-gravity forces (Figure 13-1).

Actually it's a ping-pong ball painted to look like a globe, and it's suspended using something called electromagnetic levitation. Keeping the globe in place requires careful control of the current to the electromagnet, so this project uses an Arduino interface board that is programmed from your Mac, PC, or Linux computer.

The Arduino is a great platform for this kind of situation, and the Evil Genius has used this as the basis for many projects, including that in Chapter 8 and one later in Chapter 15. Any interested Evil Geniuses may wish to check out the book *30 Arduino Projects for the Evil Genius* by the same author and publisher as this book.

What You Will Need

You will need the following components to build this project. They are listed in the Parts Bin on the next page.

Figure 13-1 The anti-gravity machine in action

PARTS BIN			
Part	**Quantity**	**Description**	**Source**
	1	Arduino Duemilanove, Uno, or clone	Internet, Farnell: 1813412
R1	1	100Ω 0.5-W metal film resistor	Farnell: 9339760
R2	1	10kΩ 0.5-W metal film resistor	Farnell: 9339787
D1	1	IR sender LED	Farnell: 1020634
D2	1	1N4004	Farnell: 9109595
T1	1	IR phototransistor	Farnell: 1045379
T2	1	N-channel power MOSFET FQP33N10	Farnell: 9845534
Coil wire	50 meters	24 AWG enameled copper wire	Farnell: 1230982
Wood (sides and top)		3 feet (1 meter) of ⅝-inch (15mm) square section	Hardware store
Wood (base)		Block about 3½" (90mm) × 1½" (40mm)	Hardware store
Wood screws	6	1½" (30mm)	Hardware store
Coil former	1	3" (80mm) × ⅜" (8mm) diameter steel bolt and nut	Hardware store
Coil washers	2	Large plastic bottle caps 1½" (35mm)	
Power supply	1	12V 1.5A (or more) power supply with 2.1mm connector	Farnell: 1279478
Magnet	2	Neodymium rare-earth magnet 10mm (⅜") in diameter	eBay, Craft and Hobby store
Ping-pong ball	1		
Small plastic bottles	2	Recycled plastic containers for a yogurt-based health drink	

The dimensions of the wood are not critical, as long as the upright sides of the frame are about 3½ inches (90mm) apart, so use whatever lumber you have on hand.

You will also need the following tools shown in the Toolbox.

Figure 13-2 shows the complete project.

TOOLBOX
■ An electric drill and assorted drill bits
■ Soldering equipment
■ A hot glue gun or epoxy resin glue
■ A wood saw
■ PVC insulating tape
■ A computer to program the Arduino
■ A USB-type A-to-B lead (as used for printers)

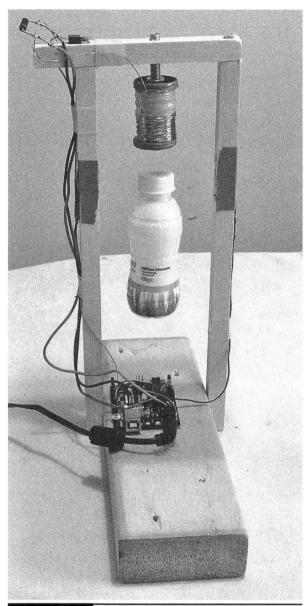

Figure 13-2 The anti-gravity machine

Assembly

The following step-by-step instructions lead you through making the anti-gravity machine. Most of the components are either soldered directly to one another or plugged into the Arduino board, so there is no need for a circuit board.

Step 1. Create the Coil Former

It's difficult to get hold of suitable coils for this kind of thing, so we are going to wind our own.

This sounds difficult, but actually it's not at all. The wire will be wrapped around a large bolt and we will use two bottle caps as the end points to contain the wire.

The first step in this is to cut the rims off the bottle caps so they are just plastic discs about 1½ inches (35mm) in diameter. We then need to drill holes in the center of each disc, the same diameter as the bolt shaft.

Fit one of the washers up against the head of the bolt and position the other washer to rest against the nut, so that the gap the coil will be wound about is 1¼ inches (30mm).

Next, wrap a single layer of tape around the bolt (Figure 13-3) to protect the enameled wire from the rough edges of the bolt's thread.

Step 2. Wind the Coil

About 700 turns of wire will be wound onto the bolt. Fortunately, we do not need to count the turns as we add them. If we fill up the space to the right depth, we should be close enough. We can also verify this by measuring the resistance of our coil, which should be between 3Ω and 5Ω.

Even better, we do not need to roll each of these turns by hand. Instead, we can use an electric drill set at its slowest speed to do it for us. Figure 13-4 shows the arrangement.

The spindle of wire is sitting on a screwdriver fixed into a vice, but it is equally practical to make

Figure 13-3 The bolt and homemade washers

Figure 13-4   Winding using an electric drill

a minion hold onto it. One end of the bolt is then secured in the drill chuck.

Before we start using the drill, we need to get the coil started.

Leave about 4 inches (100mm) of wire sticking out from the bolt at the end furthest from the head of the bolt. This will be one of our connections to the coil. Then, wind a few turns by hand to keep this loose end in place. Afterward, attach the bolt to the drill and very slowly (at first) let the wire be wrapped around the bolt. It does not matter if the coils are not laid down exactly next to each other. It is very hard to do this and it is not essential. But do try to direct the wire back and forth across the bolt in as even a manner as possible.

When the windings are about up to the top of the bottle caps, stop winding. Before cutting the

wire, we should carry out a quick resistance test to make sure we have approximately the right number of turns.

Carefully scrape away a small section of the enamel from both the lead that we started with and the wire that we were feeding onto the coil. Measure the resistance using a multimeter. It should be somewhere between 3Ω and 5Ω— ideally, 4Ω.

If the resistance is less than 3Ω, we need to add some more turns of wire to the coil. Do not worry about insulating the wire where you scraped away the enamel; the rest of the wire is insulated, so you will not cause a short circuit.

If, on the other hand, the resistance is greater than 5Ω, we need to take some turns off the coil until we bring it under 5Ω.

When you are happy with the coil, cut the wire, leaving four inches (100mm) free, and use some insulating tape to fix the wire in place (Figure 13-5).

Step 3. Build the Frame

We need a frame from which to mount the coil, Arduino board, and other components. This is easily made from four pieces of wood (Figure 13-6).

The base is made from a length of wood 1½" (90mm) × 3½" (40mm). This is really overkill, but happens to be what the author had lying around

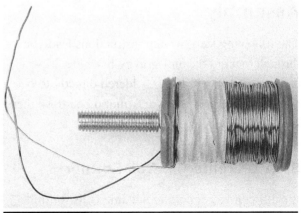

Figure 13-5   The finished coil

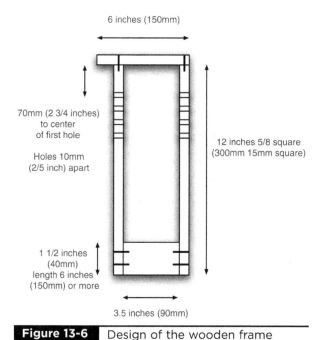

6 inches (150mm)

70mm (2 3/4 inches) to center of first hole

Holes 10mm (2/5 inch) apart

12 inches 5/8 square (300mm 15mm square)

1 1/2 inches (40mm) length 6 inches (150mm) or more

3.5 inches (90mm)

Figure 13-6   Design of the wooden frame

and does have the benefit of providing a really solid base. A smaller piece of wood is fine as long as the gap between the two uprights is about 3½" (90mm). The uprights and top piece are all secured in place with wood screws. Drill holes first with a

drill bit slightly smaller than the screws to prevent splitting of the wood.

The holes in the uprights (Figure 13-7) are 5mm in diameter so that the LEDs and phototransistor fit snugly into the holes. They should be drilled out a little with a wider bit on the inside of the uprights to remove any flakes of wood that may obscure the beam of light and to also give a slightly wider angle of view should the LED and phototransistor be misaligned.

Four holes allow for fitting the phototransistor and IR LED at different positions. The LED and phototransistor will initially be in the second hole from the top, but you may find that you need to move them up or down depending on the strength of your permanent magnet (the one in the object to be levitated) and the weight of the object.

Drill a hole that's the same diameter as the bolt. Ideally, the bolt will self-tap into the wooden top-piece so that its height can be fine-tuned by turning it. You can alternatively use a second nut on top of the cross-piece for this purpose.

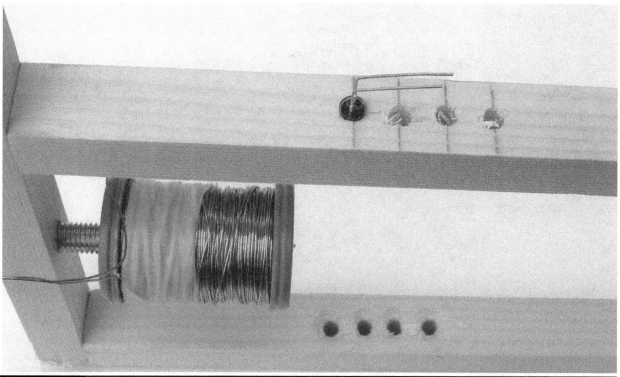

Figure 13-7   5mm holes in the uprights

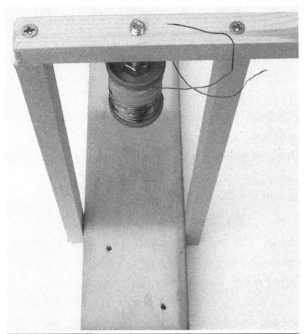

Figure 13-8 The completed frame with the coil attached

The completed frame should look something like Figure 13-8. Note that the author is no kind of carpenter!

Step 4. Mount the Transistor and Diode

We can now mount some of the electronics onto the frame. The transistor is fixed in place using one of the screws that hold the top piece of the frame in place (Figure 13-9).

Using Figures 13-9 and 13-10 as a guide, solder the negative lead of D2 (no line) to the center lead of the transistor T2. Attach the two leads from the coil to either side of the diode D2. It does not matter which lead from the coil goes to which side of the diode.

For reference, the schematic diagram is also shown in Figure 13-11.

Cut two lengths of reasonably thick wire about 12 inches (300mm) long and attach one to the positive side of D2 (bar) and one to the rightmost pin of the transistor (the source). Ideally, the first should be red and the other black. Eventually, these will be soldered to the power connector of the Arduino board.

Figure 13-9 The components on the top of the frame

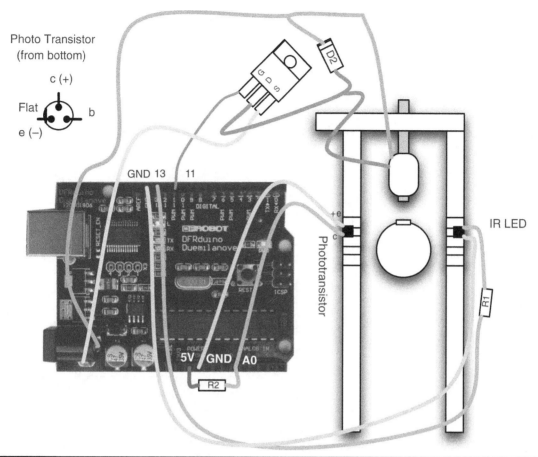

Figure 13-10   The wiring diagram for the project

Step 5. LED and Phototransistor

We will fit the LED and phototransistor into the second hole down from the top of the frame opposite each other, with the transistor on the left hand side as the frame faces us.

Cut off the third "base" lead of the phototransistor since no connection is going to be made to this. Then, bend the two remaining leads and the leads of the LED (D1) so they are at right angles to the facing direction of the LED. This way, when they are inserted into the holes, they will lie flush with the frame.

Shorten the leads of R1 (100Ω) and solder one end to the longer positive lead of the LED. Next, attach wires of about 10 inches (250mm) in length to the other end of the resistor (R1) and the unused

lead of the LED. The ends of these wires will be attached to the Arduino board.

Similarly, attach wires of the same length to the two remaining leads of the phototransistor.

Step 6. Attach the Arduino Board

We have now almost completed the electronics. All we need to do is attach all these wires to the Arduino board.

Start by soldering the power leads to the outside connections of the power socket. This is so the coil can draw current directly from the power supply without it having to go through the board. A close-up of the connections is shown in Figure 13-12.

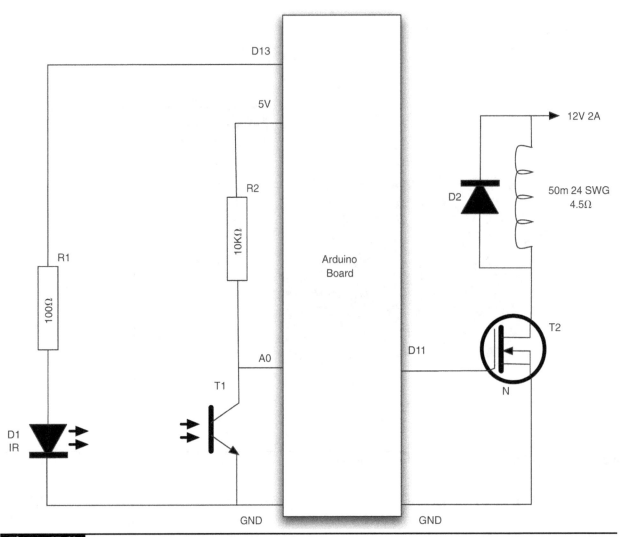

Figure 13-11 The schematic diagram for the project

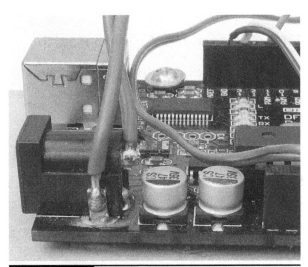

Figure 13-12 Close-up of the power connections

The other wires will all be just poked into the sockets that line the edge of the board. If you used multi-core wire, this will be easier if you tin the wires with solder first.

Starting with the leads from the LED, attach them as shown in Figure 13-10. That is, one lead to the GND socket and the other to the Digital 13 socket.

Wiring the phototransistor is a little more difficult because we are going to attach R2 at the same time. Put one end of R2 into the 5V socket and fit both the other end of R2 and the lead going to the collector of the phototransistor into the Analog In 0 socket.

Step 7. Set Up Your Computer with Arduino

To be able to program your Arduino board with the control software for the levitator, you first need to install the Arduino development environment on our computer. If you completed the persistence-of-vision project of Chapter 8, you will already have the Arduino software setup.

For installation and setup instructions, please refer back to Chapter 8 and the Arduino web site (www.arduino.cc).

Step 8. Program the Arduino Board

Without connecting the power supply to the board, connect a USB cable between your computer and the Arduino board. You should see the red power LED come on, and if it is a new board, a small LED in the middle of the board will be flashing slowly.

Launch the Arduino software. Copy and paste the code for the application, which can be viewed

at www.dangerouslymad.com, into a new project (see Figure 13-13), and then save the sketch as **antigravity**. The word "sketch" denotes a program in Arduino parlance.

To actually upload the software onto the Arduino board so it can function without the computer, click the Upload button on the toolbar. There will now be a furious flashing of little lights on the board as the program is installed on the Arduino board.

Before we make something to levitate and try the project out for real, we can carry out a few tests to make sure the sensor and LED are working okay. To do this, we will use the Arduino software's Serial Monitor. This allows your computer and the Arduino board to communicate. So you can issue commands to the Arduino, and the Arduino can also send back data.

Click the Serial Monitor (the rightmost icon on the toolbar) and after the prompt appears, type **m** into the text entry box at the top of the monitor (Figure 13-14).

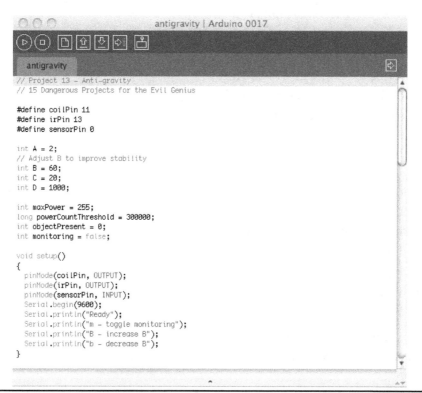

Figure 13-13 Loading the antigravity sketch

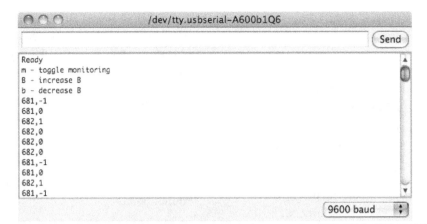

Figure 13-14 Testing with the Serial Monitor

You will then see a stream of numbers. The first number of each row is the position of the object (if any) that is being levitated, while the second number is the velocity—a positive number if the object is moving down and a negative number if it is moving up.

We do not have any object being levitated, so the position should read a high value, probably around 600. The velocity should be around 0, but may jitter up and down a bit. We can test that the transistor and LED are working okay by putting an object (say a pencil or your hand) in between them. This should decrease the "position" down to 0. You can also try moving the object up and down past the sensor and see the velocity go positive and negative as you move in different directions.

Step 9. Make Something to Levitate

The key to making an impressive distance between the electromagnet and the object being levitated is to put a strong permanent magnet at the top of the object being levitated. We use a neodymium, a rare-earth magnet. Despite their exotic-sounding name, these are easily obtainable from eBay or a hobby shop.

We could just levitate a metallic object such as a nut or bolt, but a magnet will produce much better results.

Once you have the hang of it, you can try levitating various kinds of object. For now, we will start with an easy-to-make and inherently stable object based on a very small plastic bottle that used to contain a health drink. The magnet is glued to the inside of the bottle lid, which is then screwed back onto the bottle (Figure 13-15).

Before gluing the magnet onto the lid, we need to find the right orientation for it. The best way to do this is to attach the power supply and wave a

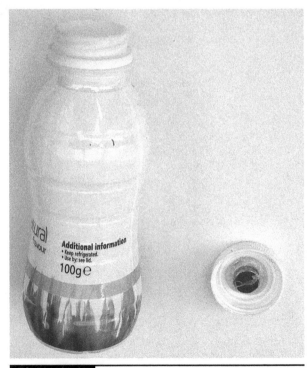

Figure 13-15 Bottle for levitation

hand in front of the beam to activate the coil. Afterward, hold the magnet first one way and then the other to find the orientation that attracts the electromagnet.

A variant on the bottle theme is to take the bottle and cut away just the top section, leaving enough to attach a ping-pong ball to it. This can then be painted to look like a globe. The end result may look something like Figure 13-16.

Step 10. Try It Out

We are now ready to try out the levitation device.

All we need to do is attach the power connector and turn it on. The program that controls the levitation device has a safety feature that turns the

coil off if there is nothing being levitated. So, to activate it, we need to move the bottle up and down a bit near the phototransistor.

If all is well, you should feel the electromagnet pull at the bottle as the bottle moves down, and relax as you move the bottle up. If you get too close to the electromagnet, you will start to feel the permanent magnet attracted to the steel bolt in the electromagnet. Therefore, you should keep the bottle out of this area.

Try to feel the forces acting on the bottle by lightly holding it from the bottom. Gradually let the bottle go and you should feel it being held by the electromagnetic force (Figure 13-17).

If there is nowhere near enough force to support the object, you may need to move the phototransistor

Figure 13-16 A globe for levitation

Figure 13-17 Levitation in action

and LED to the top hole, or move the electromagnet downward by screwing it further out of the wood.

If on the other hand the object always just flies straight up to the magnet and attaches itself to the metal bolt, you need to move the electromagnet and permanent magnet further apart.

Once you have the basic spacing about right, you may find that the bottle jumps about, rather than remaining in a stable position. If this is the case, you need to adjust parameters A, B, C, and D in the program. It is most likely to be B that needs adjusting, and fortunately you can alter this without having to change the program. Open the Serial Monitor again as if you were monitoring the position and velocity readings, but instead of entering the m command, you have two other commands. Entering a B and pressing ENTER will increase the value of B, and entering a lowercase b will decrease it. While holding onto the bottle, change the value of B until maximum stability is achieved.

Note that if you use the m option to turn on monitoring, this slows everything down, and the levitation will not work. So you cannot use the monitoring feature while simultaneously adjusting the value of B.

The Arduino board will not remember the new value of B after the power has been off, so once you have determined the best value, modify the sketch and upload it to the Arduino board again.

Theory

This project contains loads of science, from electromagnetism to Newtonian mechanics.

Basic Principles

As you will have gathered by now, the basic idea behind levitation is that as the object being levitated moves below the desired point, the power

to the electromagnet is increased, and as it rises above that point, the power is decreased.

However, using only this approach, the object will just bounce up and down, becoming more and more unstable. You can try this by changing the value of B to 0 using the Serial Monitor in the way we described earlier. To achieve any kind of stability, we must take the velocity of the object into account. If the velocity indicates that the object is above the desired position but is traveling downwards quickly, we need to ease off the power early, or the object will overshoot the target.

The formula for calculating the power is:

$$\text{Power} = \text{Position} / A + B \times \text{Velocity} + C$$

The purpose of A is to scale down the value of the position from its maximum value of about 600 to the maximum value of around 255 needed to output to the coil. So in this case, halving it is about right. If the value for the position in your arrangement is say 900 rather than 600, you should increase A to 3. Note that A is an integer value, so you cannot change it to 1.5.

Position Sensing

Figure 13-18 illustrates the way we sense the position of the item being levitated. The position is sensed by measuring the amount of IR light being obscured by the object.

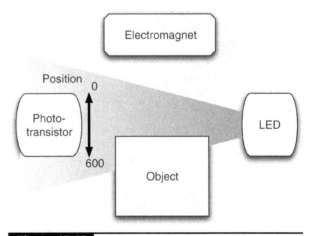

Figure 13-18 Sensing position

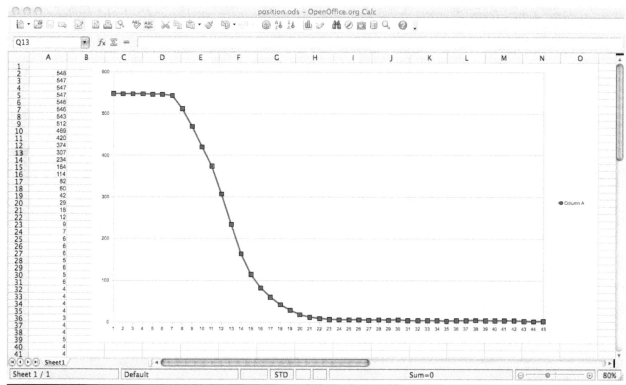

Figure 13-19 A plot of position

Figure 13-19 shows a plot of the value of the "position" variable as an obstruction (a piece of card) is gradually moved across the sensor area from bottom to top. Each step on the X axis represents one reading, while the Y axis shows the value of the "position" variable.

You can easily make your own plot like this using the Serial Monitor to capture some position and velocity readings and then paste them into a spreadsheet.

The plot clearly shows that the phototransistor starts being obscured at around sample 7 and is fully obscured by about sample 23. This means that over the 5mm distance range that the sensor covers, we can resolve about 550 different values, which is pretty accurate.

Ambient Light Compensation

A light sensor for determining the position of the item being levitated can suffer from problems with changes in ambient light. That is, if someone turns the lights on or the sun comes out, more light would enter the phototransistor, giving a false reading for the position.

This anti-gravity machine uses a cunning technique to compensate for such changes. It momentarily turns off the LED, takes a measurement of the ambient light, and then turns the LED back on again. All of this within a few thousandths of a second. The software for doing this is described in the next section.

Figure 13-20 shows a plot of the variables in the program for "raw" light intensity, ambient light intensity. and the difference between the two. The first two of these are captured using the Serial Monitor with a modification to the software to write out the values of the "raw" and "ambient" variables.

The first thing to notice is that the raw value is considerably higher than the ambient value. This

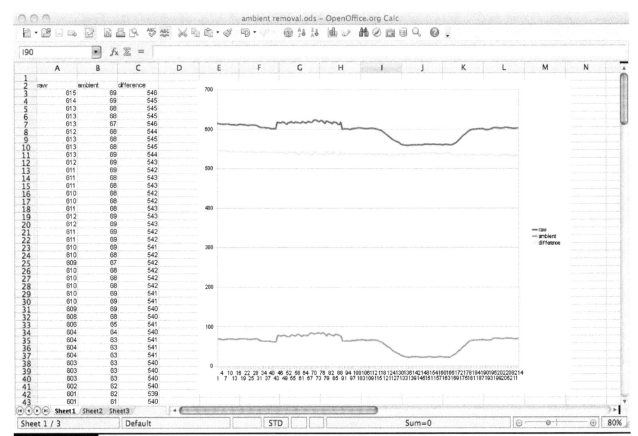

Figure 13-20   A chart showing ambient light compensation

means that the IR LED is doing its work and shining a bright beam into the phototransistor, so we are naturally keeping the influence of external light fairly low.

The difference line remains fairly flat, which is what we want, and means that external influences such as turning on a bright light near the levitator (between samples 43 and 90) and shading it with a large box (between samples 118 and 180) were both compensated for.

Interestingly, you can see the mains frequency flicker from that first event, where the brightness of the lamp fluctuated with the mains frequency. This leaves a small trace on the compensated signal, but not enough to worry about.

Control Software

If you have not used Arduino before, then hopefully you will be impressed by the simplicity and power of this little microcontroller board. Let's have a look at the program that controls the levitation machine. (See Listing 13-1.)

The Arduino boards use the C programming language. It is beyond the scope of this book to explain the whole C language, but if you have a basic familiarity with programming, you should be able to understand the following explanation.

The program starts with some variable declarations. The lines that start with a // denote comment lines that explain what is happening.

We then have a "setup" function that says how the various connections on the board are to be used, and sets the baud rate for communication with the Serial Monitor. A message saying what

LISTING 13-1

```
// Project 13 - Anti-gravity
// 15 Dangerous Projects for the Evil Genius

#define coilPin 11
#define irPin 13
#define sensorPin 0

int A = 2;
// Adjust B to improve stability
int B = 60;
int C = 20;
int D = 1000;

int maxPower = 255;
long powerCountThreshold = 300000;
int objectPresent = 0;
int monitoring = false;

void setup()
{
  pinMode(coilPin, OUTPUT);
  pinMode(irPin, OUTPUT);
  pinMode(sensorPin, INPUT);
  Serial.begin(9600);
  Serial.println("Ready");
  Serial.println("m - toggle monitoring");
  Serial.println("B - increase B");
  Serial.println("b - decrease B");
}

void loop()
{
  static int count = 0;
  static int oldPosition = 0;
  static int ambient = 0;
  static long powerCount = 0;
  count ++;

  if (count == 1000)
  {
    ambient = readAmbient();
    count = 0;
    objectPresent = (powerCount < powerCountThreshold);
    powerCount = 0;
  }
  int raw = 1024 - analogRead(sensorPin);
  // position from top (0) of sensor region to the bottom (650)
  int position = raw - ambient;
  // positive value means going downwards, negative going upwards
  int velocity = position - oldPosition;
  int power = position / A + velocity * B + C;
```

(continued on next page)

LISTING 13-1 *(continued)*

```
  powerCount += power;
  oldPosition = position;

  // clip
  if (power > maxPower) power = maxPower;
  if (power < 0) power = 0;

  checkSerial();

  if (monitoring)
  {
    Serial.print(position);  Serial.print(",");
    Serial.println(velocity);
  }

  analogWrite(coilPin, power * objectPresent);
  delayMicroseconds(D);
}

int readAmbient()
{
  digitalWrite(irPin, LOW);
   // allow time for LED and phototransistor to settle
  delayMicroseconds(100);
  int ambient = 1024 - analogRead(sensorPin);
  digitalWrite(irPin, HIGH);
  return ambient;
}

void checkSerial()
{
  if (Serial.available())
  {
    char ch = Serial.read();
    if (ch == 'm')
    {
      monitoring = ! monitoring;
    }
    if (ch == 'B')
    {
      B += 5;
      Serial.println(B);
    }
    if (ch == 'b')
    {
      B -= 5;
      Serial.println(B);
    }
  }
}
```

commands can be issued is then echoed to the Serial Monitor.

The "loop" function is automatically run repeatedly. It is this loop that controls the coil.

First, we define some static variables that will retain their value between calls of the function. The variable "count" is incremented each time "loop" is run; "oldPosition" is used to find the change in position between samples, and hence the velocity; "ambient" is used to contain the ambient light intensity when the LED is off; and, finally, "powerCount" is used to keep track of how long the coil has been on and whether anything is being levitated or not.

The "if" statement allows the code inside it to be run just one time in every thousand. The first thing it does is measure the ambient light—that is, the light reading you get when the LED is off. This is then used to adjust the light reading so the levitation device is not affected by changes to the lighting when a light is turned on. This happens because the "readAmbient" function momentarily turns off the LED and takes a reading of the light before turning the LED back on again.

The next part of the one-in-a-thousand block sets the "objectPresent" variable depending on the accumulated "powerCount." This is important because it allows the coil to be turned off when there is no object being levitated. If we did not do this, then maximum power would be flowing through the coil whenever there was nothing being levitated. This would make the coil dangerously hot after a while.

After the "if" statement, we do the calculations to find the position and velocity, and then calculate the power we should apply to the coil. If the power is outside the range for the analog output, then the next few lines "clip it."

The "checkSerial" function checks for any incoming commands from the Serial Monitor and processes them. If the m command is issued, then the "monitoring" variable is set. Back in the "loop" function, if "monitoring" is true, then the position and speed will be written to the Serial Monitor. While this is happening, it will slow everything down, likely causing levitation to fail.

Finally, we write the power level to the analog output controlling the power, and then delay for a period of time determined by parameter D before beginning the whole process again.

Summary

This is one of the easier projects in the book to build, but one of the most difficult to get right. It will probably take some experimentation on your part to make it all work. But the first time you get something to levitate, it all becomes more than worthwhile.

Feel free to experiment with levitating other items—even items without a magnet in them. To do this, however, you will need to move the sensor closer to the electromagnet.

In the remaining two chapters of this book, we switch our attention to two very different kinds of robots. Next stop: a tiny light-seeking robot that uses toothbrush motors.

CHAPTER 14

Light-Seeking Microbot

<table>
<tr><td>PROJECT SIZE:</td><td>Small</td></tr>
<tr><td>SKILL LEVEL:</td><td>★★★★</td></tr>
</table>

SOMETIMES THE EVIL GENIUS GROWS tired of the company of his minions and needs something more stimulating and intelligent. At these times, he turns his attention to a little electronic companion: "Snailbot."

As you can see, when compared to a AA battery (as shown in Figure 14-1), this robot is very small.

This tiny snail will slither towards a flashlight held in front of it—this in contrast to minions who will generally just freeze in terror when a light is shone in their eyes.

The project uses a small number of components and is simple to construct. It employs the recycled electric motors from disposable toothbrushes to drive the robot, and its power is supplied from a tiny snail shell–like rechargeable battery.

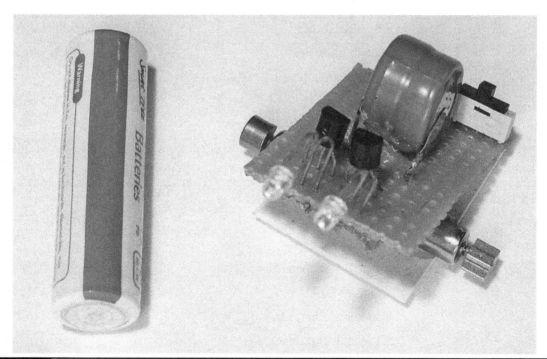

Figure 14-1   Snailbot

What You Will Need

You will need the following components to build this project. They are listing in the Parts Bin.

The last seven components are only required if you need to make a small charger for the robot. If you have a variable current power supply, you don't need to build this part of the project.

You will also need the following tools:

TOOLBOX
■ Soldering equipment
■ A hot glue gun or self-adhesive pads
■ A hacksaw
■ A bright flashlight

Assembly (Robot)

Figure 14-2 shows the schematic diagram for the robot. See the "Theory" section at the end of this chapter for more information about how this circuit works. The basic idea is that the phototransistors control the power to the motors. The more light falling on the left phototransistor, the more power goes to the right motor. This "cross-over" arrangement will naturally cause the robot to home in on a light source.

The following step-by-step instructions lead you through making the robot. Everything apart from the motors is built onto a small piece of stripboard.

Step 1. Charge the Battery

We will not be able to test Snailbot if its battery is empty. So we should first charge it up.

If you have a bench power supply with variable current, then set the current to 10mA, clip the leads to the battery's terminals (positive to positive), and let it charge for ten hours.

If you don't have one, then you need to read the later section in this chapter called "Assembly (Charger)" and build the battery charger first.

PARTS BIN			
Part	**Quantity**	**Description**	**Source**
T1, T3	2	Phototransistor	Farnell: 1497673
T2, T4	2	BC548 transistor	Farnell: 1467872
Motors	2	Vibrating motor from toothbrush	
S1	1	PCB switch	Farnell: 674345
Battery	1	80mAh 2.4V MiMH battery	Farnell: 854657
Stripboard	1	Stripboard; 11 strips, each with 14 holes	Farnell: 1201473
Skid	1	Small strip of plastic 1¼" (33mm) × 1" (25mm)	
For the Charger			
Battery clip	1	PP3-style battery clip	Farnell: 1183124
Battery	1	PP3 battery	
R1	1	470Ω 0.5-W metal film resistor	Farnell: 9338810
D1	1	5mm red LED	Farnell: 1712786
Clip	1	Crocodile clip (red)	Farnell: 1169601
Clip	1	Crocodile clip (black)	Farnell: 1169604
Wire		Short lengths of red and black wire	

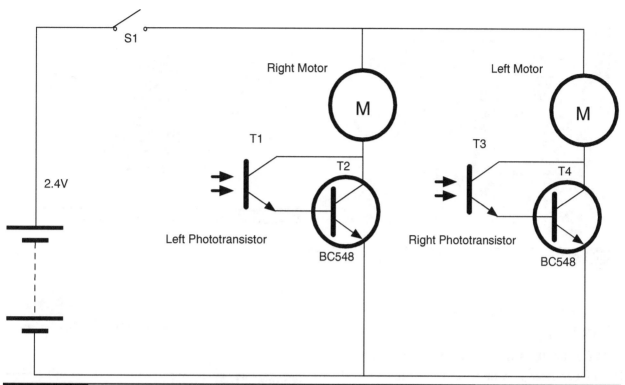

Figure 14-2 The schematic diagram

Step 2. The Stripboard

Figure 14-3 shows the stripboard layout.

Cut the stripboard into an area of 11 strips, each with 14 holes. You can do this with a strong pair of scissors or by scoring both sides of the board with a craft knife and then breaking it over the edge of a table.

There is only one linking wire to fit to the stripboard, and this should be soldered in place first (Figure 14-4).

Next, solder in the switch and the battery. The negative lead of the battery is attached to the link, and the positive lead to a small wire passing through the board to the track below. The battery

Figure 14-3 The stripboard layout

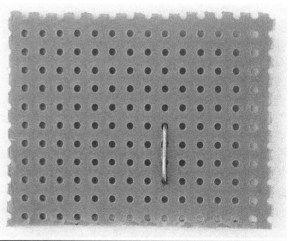

Figure 14-4 The stripboard with the link in place

Figure 14-5   The switch and battery in place

Figure 14-6   The completed board

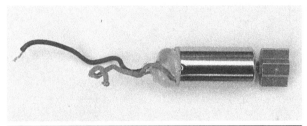

Figure 14-7   Toothbrush motor

should be positioned as centrally as possible (Figure 14-5).

Make sure the switch is in the off position. Remember, from now on we are soldering a board that has a charged battery attached to it, so it would be quite easy to accidentally short the battery while soldering, which would likely damage it.

The transistors can now be soldered into place. The phototransistors are soldered in place, leaving their leads as long as possible to allow them to be bent into a snail-eye's position (Figure 14-6).

Step 3. Salvage the Motors

The Evil Genius is a considerate employer and buys low-cost disposable electric toothbrushes for his minions to use when they are cleaning his teeth. Throwing these away, just because the bristles are a little frayed, seems very wasteful to any Evil Genius who likes to do a bit of recycling.

These toothbrushes contain tiny lightweight electric motors (Figure 14-7) with a semicircular weight that makes the whole thing vibrate when the toothbrush is turned on. A typical example of such a brush is shown in Figure 14-8.

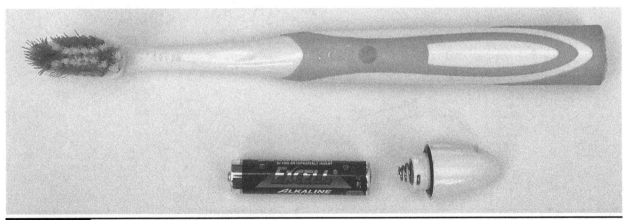

Figure 14-8   Disposable electric toothbrush

You should select toothbrushes that are of a thin design and are powered by a single AAA battery.

Extracting the motor requires a good deal of care, as it is all too easy to break off the leads that connect to the motor.

The first step is to remove the battery and then chop out the middle section from the switch assembly to the beginning of the brush. You can just hacksaw straight through, leaving a middle section.

Now comes the delicate bit. Figure 14-9 shows this middle section carefully sawn open along its length and the switch and motor assembly removed. The motor can be de-soldered and the rest of the toothbrush discarded.

Step 4. Attach the Motors

The two motors are attached to a strip of plastic 1 inch (25mm) wide (Figure 14-10). Before gluing the motors into place, bend the plastic slightly so it forms a gentle arc. This will let Snailbot slide across surfaces more easily.

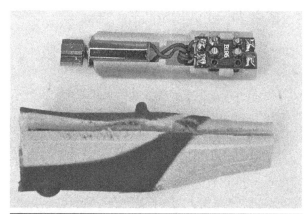

Figure 14-9 Removing the switch and motor assembly

The weights on the motors are not going to be in direct firm contact with the ground, but rather the vibrations and occasional glancing contact by the weights on the motors with the ground will propel Snailbot forwards.

Notice that in Figure 14-10 one of the motor leads has been extended. You may well need to do this so the leads will easily reach their points of connection on the underside of the stripboard.

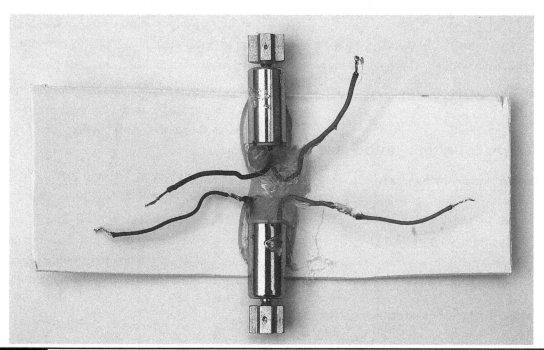

Figure 14-10 Attaching the motors

Step 5. Final Wiring and Test

Using Figure 14-3 as a guide, solder the leads of the motors to the underside of the stripboard.

We can now carry out a test to make sure everything is right before we glue it all together.

Flip the switch to the "on" position. Both motors should be still. If either or both of them are turning a little, put your hand over the phototransistors and the motors should stop. This just means your workbench is getting too much illumination, so find a darker place.

Now take a flashlight and shine it in Snailbot's left eye. This should make the right motor whir. It should whir faster the closer you move the flashlight to it. The other motor will probably pick up some of the stray light and spin a little, too. You should, however, be able to tell that each phototransistor controls the opposite motor.

Finally, check that the motors are turning the right way to propel Snailbot forwards. If they are not, you may be able to turn the board through 180 degrees without having to re-solder the connections. If the leads are not long enough to do this, you will have to re-solder all the leads, switching the red and blue leads around for each motor.

If one of the motors is not working, go back and carefully check the wiring and make sure the phototransistor for that motor is the right way around.

Step 6. Glue Everything into Place

The stripboard is glued by its rear edge to the plastic so it is more or less centered over the motors (Figure 14-11). The excess plastic can then be cut off to the right length with a pair of scissors.

Using Snailbot

Snailbot works best on a table top or other flat surface. It does not work so well on carpet. Take a

Figure 14-11   Fixing the stripboard to the plastic

flashlight and shine it directly in the face of Snailbot and it should whir to life, chasing after the flashlight as you move it round the table top.

Assembly (Charger)

If you have a variable current power supply, you probably do not need to make this charger. It charges Snailbot from a 9V battery and is shown in Figure 14-12.

Figure 14-13 shows the schematic diagram for Snailbot's charger, and as you can see, it is a very simple design, comprised of a battery clip, an LED, a resistor, and two crocodile clips.

The charger has few enough components that you can just solder all the components to each other, as shown in the wiring diagram of Figure 14-14.

To use the charger, first turn the switch off. The crocodile clips attach directly to the battery.

This charger operates at about 10mA, which means a full charge will require an overnight charging of about ten hours. But once charged, you should find you get hours of fun from your Snailbot.

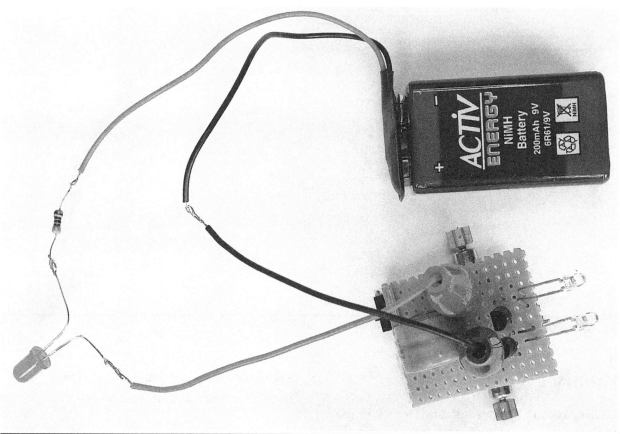

Figure 14-12 A simple charger for Snailbot

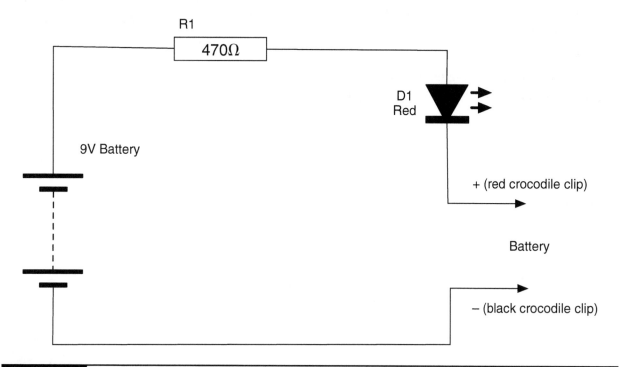

Figure 14-13 The schematic diagram for Snailbot's charger

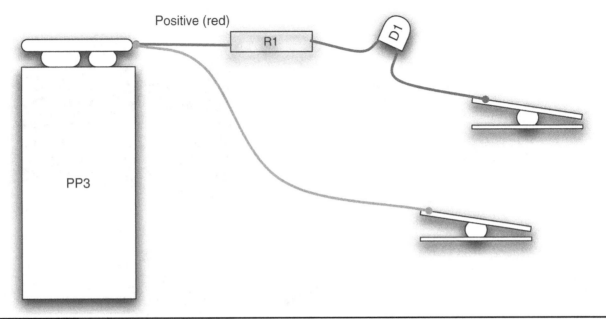

Figure 14-14 Snailbot charger wiring diagram

Theory

Refering back to Figure 14-2, this is a very simple design.

The phototransistors do not have sufficient amplification to drive the motors directly and so each phototransistor has a second transistor connected to it in an arrangement called a Darlington-Pair.

The smaller current flowing through the collector of the phototransistor is fed directly into the base of the second transistor, providing "double" amplification of the current.

Summary

This is a small and very simple robot. In the next and final chapter, we will scale up to make a much bigger and more capable robot—in fact, a robot with a brain, albeit a small microcontroller-based brain.

CHAPTER 15

Surveillance Robot

PROJECT SIZE: Large

SKILL LEVEL: ★★★★

THE EVIL GENIUS' LAIR IS PATROLLED by a security robot. This robot (Figure 15-1) wanders around the lair in a random manner, occasionally tripping up minions. It also serves the more practical but less amusing role of monitoring the lair when the Evil Genius is asleep at night.

The robot is equipped with a passive infrared sensor and a loud alarm. It will move around, occasionally stopping and monitoring the area in front of it with its PIR movement sensor. If it then detects any movement, it sounds the alarm.

This is another project based on an Arduino microcontroller board. It uses modified low-cost electric screwdrivers to provide powerful geared motors to drive it along, and an infrared distance sensor to prevent collisions (although this is not always successful).

If you want to learn a bit more about robotics, this project provides an excellent starting point, and you can extend the project in several ways—for example, by adding a wireless camera and remote control.

Assembly

This is a complex project, so to simplify matters we are going to divide it up into a number of modules, each of which can be assembled and tested separately. The overall arrangement of the modules is shown in Figure 15-2.

You can see that we are going to use two batteries: one 9V PP3 battery to power the Arduino and other components such as the PIR sensor that require 9V, and a second battery comprised of six rechargeable AA batteries in a battery holder. This battery supplies the power to the electric motors.

We will start by building the motor controllers, and then build the rest of the electronics into and onto a project box.

Figure 15-1   Surveillance robot

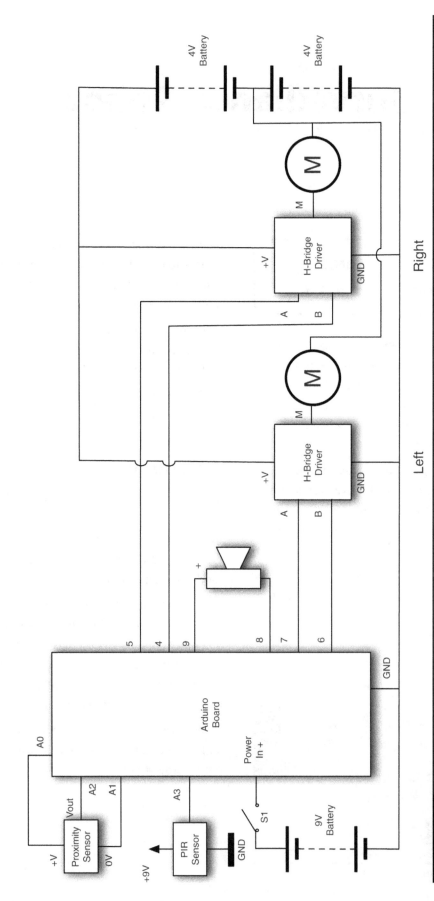

Figure 15-2 Schematic diagram for the surveillance robot

Assembling the Motor Drivers

These motor drivers are useful little modules that you may well find you can use in other projects. They are easy to construct and do not require many components.

What You Will Need

You will need the components that are listed in the Parts Bin to build each motor controller. Since we need two controllers, the quantities have been doubled. You will also need the six AA batteries, the holder, and the clip to be able to test the controller.

You will also need the following tools:

TOOLBOX
■ Soldering equipment
■ A 6V motor or Rtest (see component list)
■ A multimeter

The schematic diagram for the motor controller is shown in Figure 15-3.

The design uses an arrangement of two MOSFET transistors so the motor can be powered in either direction, depending on the control pins A and B. For a description of motor control designs, see the "Theory" section at the end of this chapter.

Step 1. Prepare the Stripboard

Figure 15-4 shows the stripboard layout for one of the motor driver modules.

The first step is to cut a piece of stripboard that has 12 strips, each with 12 holes. Two breaks must be made in the tracks. Make these using a drill bit as a hand tool, rotating it between your finger and thumb. Figure 15-5 shows the prepared board ready for the components to be soldered to it.

PARTS BIN			
Part	**Quantity**	**Description**	**Source**
R1–R3	6	10kΩ 0.5-W metal film resistor	Farnell: 9339787
R4	2	1kΩ 0.5-W metal film resistor	Farnell: 9339779
Rtest (optional)	2	100Ω 0.5-W metal film resistor	Farnell: 9339760
T1	2	NPN bipolar transistor BC548	Farnell: 1467872
T2	2	P-channel power MOSFET FQP27P06	Farnell: 9846530
T3	2	N-channel power MOSFET FQP33N10	Farnell: 9845534
Terminal (power)	2	Four-way 5mm-pitch PCB terminal block	Farnell: 9632697
Terminal (control)	2	Two-way 5mm-pitch PCB terminal block	Farnell: 9632670
Stripboard	1	Stripboard—this is enough for both motor controllers	Farnell: 1201473
Batteries	6	AA rechargeable batteries	
Battery holder	1	Battery holder for 6 × AA batteries	Farnell: 3829571
Battery clip	1	Battery clip for 9V PP3 battery	Farnell: 1650667

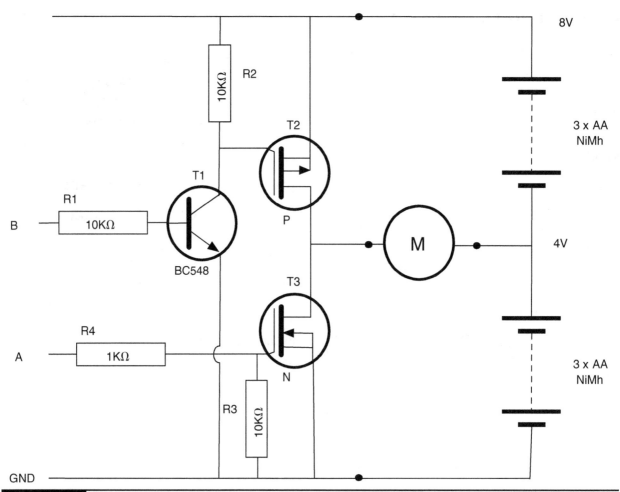

Figure 15-3   Schematic diagram for the motor controller

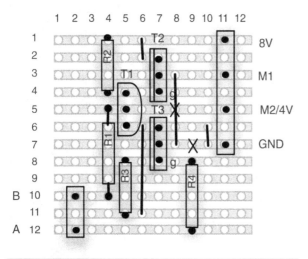

Figure 15-4   The stripboard layout for a motor driver module

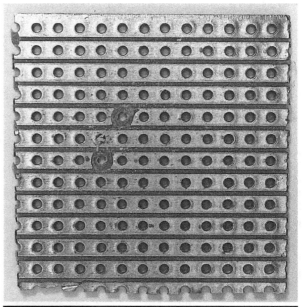

Figure 15-5   A motor driver module stripboard ready for soldering

Step 2. Solder the Links and Resistors

Before starting on the real components, solder the four wire links into place.

We can now solder all the resistors in place. Once done, your board will look like Figure 15-6.

Step 3. Solder the Remaining Components

Since the terminal blocks are slightly lower in profile than the transistors and capacitor, solder them into place next.

Place the transistors into position, ensuring they are the correct way around and that the P-channel transistor is at the top of the board. Then, solder each transistor into place.

When all the components are soldered, the board should look like Figure 15-7.

Figure 15-7 The board with all components in place

Step 4. Test the Motor Controller Module

To test the motor controller, set up the arrangement shown in Figure 15-8. You will need to attach a lead to the 4V point in the battery holder. See Step 9 of the next section if you are not sure how to do this. You can either use the motor and see it rotate both one way and the other, or use the test resistor and just measure the voltage across it.

The lead connected to +8V will momentarily touch the screw of either A or B, but never both at the same time.

Initially, make sure there are no connections to either A or B. This will ensure that both transistors are turned off.

Touch the loose lead from +V to A, causing the motor to turn one way. The meter should read –4V. Touch it to B and it should read +4V.

The effect of the control pins A and B are summarized in Table 15-1.

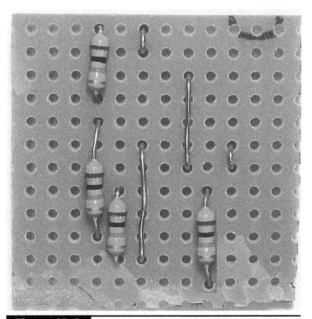

Figure 15-6 The board with resistors and links

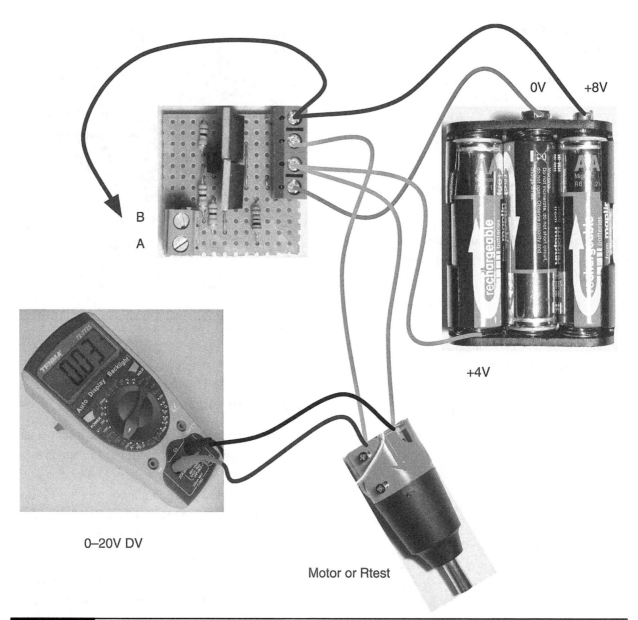

Figure 15-8 Testing a motor controller module

TABLE 15-1	Results of Pin A and B	
A	**B**	**Result**
GND or no connection	GND or no connection	Motor stopped
+V	GND or no connection	Motor turns clockwise
GND or no connection	+V	Motor turns counterclockwise
+V	+V	BANG!

The "BANG!" condition is both interesting and important. What it signifies is that if both A and B are connected to +V at the same time, then both MOSFET transistors will be on at the same time, meaning there is effectively a short circuit between 8V and ground. This will result in very large currents flowing through the transistors, which will probably result in smoke and destruction.

This is something we must keep at the fore of our thinking when it comes time to write the software for the Arduino.

So now we have a working motor controller. But we need two, so repeat this whole process again.

Assembling It All

Now that we have a pair of working motor controllers, we need to move on to assemble the rest of the robot.

Figure 15-9 shows the wiring diagram for the project.

What You Will Need

This section assumes we already have two assembled motor controllers. The other components we need for the rest of the project are listed in the following Parts Bin.

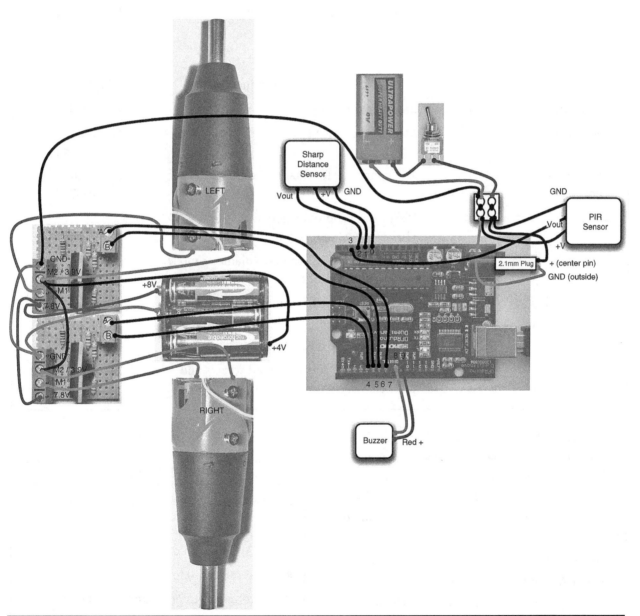

Figure 15-9   Wiring diagram for the project

		PARTS BIN	
Part	**Quantity**	**Description**	**Source**
Pin headers	1	Pin header strip, broken into two sections of four pins, and one section of two pins	Farnell: 1097954
Electric screwdrivers	2	Budget 6V electric screwdriver using 4 × AA batteries	eBay
Arduino	1	Arduino Uno or Duemilanove	Internet, Farnell: 1813412
Terminal block	1	Two-way 2 A screw terminal block	Farnell: 1055837
Switch	1	SPST miniature toggle switch	Farnell: 1661841
Battery clip	1	PP3 9V battery clip	Farnell: 1650667
Battery	1	PP3 battery	
PIR	1	Miniature PIR sensor 3 terminal 9–12V	eBay
Power connector	1	2.1mm power plug	Farnell: 1200147
Distance sensor	1	Sharp IR distance sensor (GP2Y0A21YK)	Sparkfun: SEN-00242
Distance sensor connector	1 (optional)	JST connector for Sharp IR distance sensor	Sparkfun: SEN-08733
Case	1	Plastic case approximately 6½" × 3½" × 2½" (165mm × 85mm × 55mm)	
Big wheels	2	Approximately 2½" (65mm) diameter model car wheels	Model/toy shop
Small wheels	2	Approximately 1½" (35mm) diameter model car wheels	Model/toy shop
Bracket	2	Plastic, cut into an L shape	
Axle	1	Short metal axle to suit small wheels and plastic brackets	Model/toy shop

An older Arduino Duemilanove is absolutely fine for this project. However, if you want to use the latest kit, get an Arduino Uno. In either case, clones of the open-source hardware are available for considerably less money than original Arduino boards.

The screwdrivers should be of the type that takes four AA cells. The author found these available for around $8 each, made by Duratool.

The wheels were sourced from a toy shop.

You will also need the following tools:

TOOLBOX
■ An electric drill and assorted drill bits
■ A hacksaw
■ A hot glue gun or epoxy glue
■ Assorted self-tapping screws
■ A computer to program the Arduino
■ A USB-type A-to-B lead (as used for printers)

Step 1. Disassemble the Screwdrivers

The first step is to remove the switch and battery holder end of the screwdrivers, since all we are really interested in is the motor and gearing. Figure 15-10 shows one of the screwdrivers with the battery case sawn off.

To find the right place to saw, remove the end-cap and battery holder. Then, saw where you think they would end were they in the case. If in doubt, saw off less rather than more. You can always saw more off in another cut.

You need to expose the terminals to the motor. Once exposed, solder leads two or three inches in length to the terminals.

Step 2. Make the Large Wheels

One of the nice things about using electric screwdrivers is that we can use the screwdriver bits that they come with to attach the wheels. The screwdriver that the author used was magnetic, so the screwdriver bits could be glued to the big wheel and then the wheels could be attached to the screwdriver as if they were bits (Figure 15-11).

Figure 15-11 The large wheel

Step 3. Fix the Motors to the Case

The motors are fixed to the bottom of the case using self-tapping screws from the inside of the case, which are then fitted into the body of the screwdriver. Be careful to select screws that are long enough to go into the screwdriver casing without sticking into anything vital, like the motor casing.

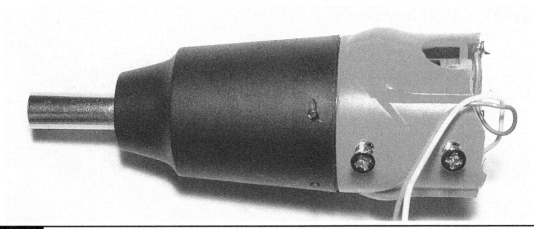

Figure 15-10 Sawn-off screwdriver

Figures 15-12 and 15-13 show the screwdrivers attached from the inside and the outside. Note also the extra holes to allow the leads from the motors to enter the project box.

Step 4. Attach the Small Wheels

The two front wheel brackets are made by bending strips of plastic salvaged from packing material. They are then drilled at the end of the long end to take the axle, and at the short end to take self-tapping screws to attach them to the bottom of the project box.

Figure 15-14 shows the wheels attached to the box.

The front wheels actually do very little besides stopping the robot from falling over. Do not worry if they will not turn easily; when the robot is rotating on the back wheels, they will be traveling sideways.

For this reason, it's actually a good idea for the front wheels to be smooth and not have a good grip.

Step 5. Attach the PIR Sensor

The PIR sensor (Figure 15-15) is designed for use in intruder alarms. The unit has three pins: +V, GND, and Vout. They are designed to work at 12V, but most will work with a supply of 9V. In this unit, the output was normally at 0V, but when movement was detected, the output rose to around 3V for about a second.

To test your PIR sensor, temporarily attach the battery clip and PP3 to the GND and +V screw terminals on the PIR sensor, then use a multimeter to measure Vout for the resting case and when you

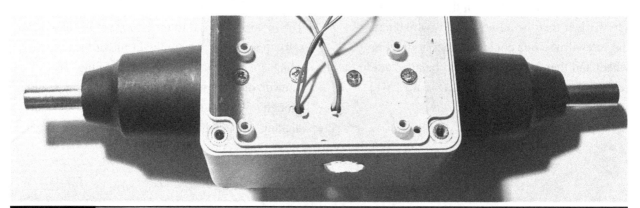

Figure 15-12   The electric screwdriver motors—from the inside

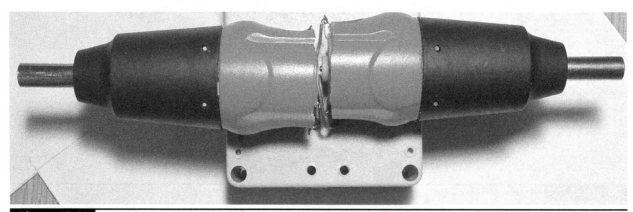

Figure 15-13   The electric screwdriver motors—from the outside

Figure 15-14 Attaching the front wheels to the box

wave your hand in front of it. Chances are your sensor will behave the same; however, it may produce a different output when detecting movement. If this is the case, you will have to alter the script, in particular, the variable pirThreshold. The unit of this variable is roughly the voltage * 200. So a value of 500 means that

the alarm will be triggered if the voltage goes over 500/200 or 2.5V.

Attach three wires to the PIR sensor terminals and thread them through a hole in the lid of the box, then glue the sensor to the box, facing forward (Figure 15-16).

Step 6. Attach the Buzzer

The buzzer used had red and black trailing leads for its positive and negative connections. It is attached in a similar way to the PIR sensor with a small blob of glue, and the leads are fed through a hole in the lid (Figure 15-16)

Step 7. Attach the Proximity Sensor

The IR proximity detector (Figure 15-17) is a clever little device that uses reflected IR light to determine its distance from any obstacle. It will measure objects ranging from 100 to 800mm.

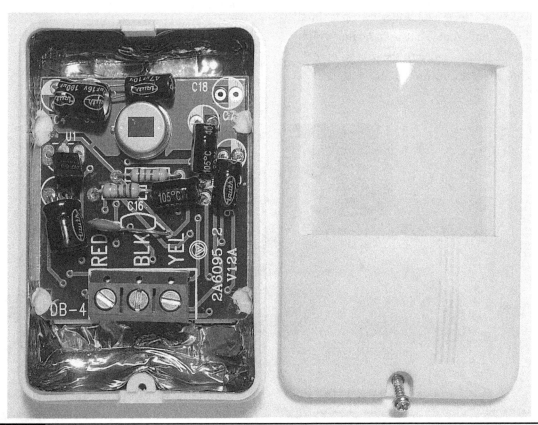

Figure 15-15 The PIR sensor

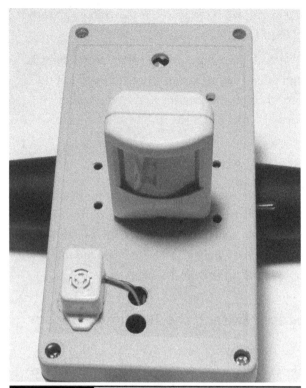

Figure 15-16 The PIR sensor and buzzer attached to the box lid

Rather like the PIR sensor, the sensor has three pins: GND, 5V, and Vout. Vout is an analog voltage that is related to the distance to the sensed object. We only care if we are going to bump into

something, so all we need to know is that if the voltage at Vout rises above 2V, then we are within 120mm of hitting something, and we need to stop.

The IR sensor is attached to the front of the project box using two small self-tapping screws.

Step 8. Attach Pin Headers to the Ribbon Cable

To make connections to the Arduino, we use pin headers and short lengths of ribbon cable scavenged from an IDE hard disk cable.

Using Figure 15-9 as a reference:

- Solder the two-way header pin to the end of the buzzer leads.

- Solder the three leads from the distance sensor to three of the leads from one of the four-way headers.

- Solder the Vout lead from the PIR sensor to the remaining free pin of the four-way header used for the distance sensor.

- Solder four leads to link both motor controllers' control pins A and B to the other four-way header.

Figure 15-17 IR proximity sensor

Step 9. The Switch and Batteries

Still referring to Figure 15-9, cut the red lead on one of the battery clips and solder the switch to one side of it. The black negative lead of the battery clip and the lead from the other side of the switch are connected to one side of the two-way terminal block. Attach leads to the 2.1mm power connector and fit the leads into the other side of the terminal block, each accompanied by one of the GND and V+ connections to the PIR sensor. The 2.1mm plug can then fit into the Arduino.

Note, that we only switch the power to the 9V battery. The motor controllers will effectively keep the six-cell motor battery disconnected.

The battery for the motors needs the normal V+ and GND supplied by a second battery clip, but also requires a center supply. We will place a wire under one of the sprung connections and then use the battery to hold it in place.

Figure 15-18 shows the motor battery with the extra lead attached.

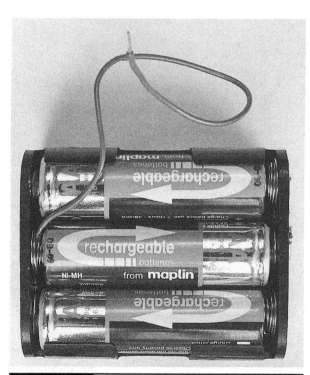

Figure 15-18 The motor battery

To find the right place, we are going to use a multimeter. The procedure is to first measure the full voltage across the connectors for the battery clip. This will probably be a touch under 8V. Leave the negative test lead connected to the battery negative and work your way around the connectors for the cells until you find a sprung one at about 4V. You can then tuck a wire into the sprung connector.

We can now wire up the leads from the motor to the terminal blocks on the motor controller, as well as the other connections from the motor controllers to the battery and GND from the other battery.

Finally, we can plug the headers into the Arduino board. Make sure they are the correct way around.

Do not plug the motor controller connector into the Arduino board until after the board is programmed, as you do not know the state of the output pins, and there may be the possibility of A and B both being high at the same time for one of the motor controllers.

Many connections need to be made here, so it may be worth checking each connection on Figure 15-9 as it is made to make sure you don't miss one.

Step 10. Program the Arduino

Our robot is not going to do much until we program it. If you have not read about the projects in Chapters 8 and 13, which also use an Arduino, I strongly recommend you read the relevant parts of Chapter 8. This gives some background on the Arduino and explains how to install and set up your computer with the Arduino software.

The sketch (program) for the robot can be found at www.dangerouslymad.com. Paste it into a new Arduino project and upload it to the Arduino board using the instructions in Chapter 8.

Step 11. Testing the Robot

Set it down in the middle of the room and turn it on. It will start in its monitoring mode, so if you move in front of it, it should detect you and sound the alarm. After a while, it should rotate in a random direction and set off for its new position.

If it finds itself on a collision course, it will back up a little before going back into monitoring mode.

You may find that your motor connections are the wrong way around and have to swap over a few leads. If your robot is behaving erratically, you can find a helpful test sketch on this at www.dangerouslymad.com. It lets you send commands using the Serial Monitor to move forward, back, left, and right, as well as report the values from the sensors.

Theory

In this section, we take a detailed look at the code for the robot, the distance sensor and also look at techniques for controlling motors.

The Program

The sketch (Arduino parlance for a program) is shown in Listing 15-1.

The program makes a great starting point for anyone wishing to develop robots using the Arduino.

At the top of the file, we define the various pins to be used. The two variables—proxThreshold and pirThreshold—set the value that must be exceeded to indicate an imminent collision or detection of movement, respectively, using the outputs of these

---

LISTING 15-1

```
// Surveillance Bot

#define HALT 0
#define CLOCKWISE 1
#define COUNTER_CLOCKWISE 2

int leftAPin = 7;
int leftBPin = 6;
int rightAPin = 5;
int rightBPin = 4;

int posPin = 14;
int negPin = 15;
int proxPin = 2;
int pirPin = 3;
int buzzPlusPin = 9;
int buzzMinusPin = 8;

float proxThreshold = 500;
float alpha = 0.5;
int pirThreshold = 10;
int monitorDuration = 120; // seconds
int alarmDuration = 10; // seconds

void setup()
{
  pinMode(leftAPin, OUTPUT);
```

LISTING 15-1   *(continued)*

```
  pinMode(leftBPin, OUTPUT);
  pinMode(rightAPin, OUTPUT);
  pinMode(rightBPin, OUTPUT);
  pinMode(pirPin, INPUT);

  digitalWrite(leftAPin, LOW);
  digitalWrite(leftBPin, LOW);
  digitalWrite(rightAPin, LOW);
  digitalWrite(rightBPin, LOW);

  pinMode(posPin, OUTPUT);
  pinMode(negPin, OUTPUT);
  pinMode(buzzPlusPin, OUTPUT);
  pinMode(buzzMinusPin, OUTPUT);
  digitalWrite(posPin, HIGH);
  digitalWrite(negPin, LOW);
  Serial.begin(9600);
}

void loop()
{
  monitor();
  moveToNewPlace();
  delay(1000);
}

void monitor()
```

LISTING 15-1 *(continued)*

```
{
  int alarmTimeout = 0;
  for (int i = 1; i < monitorDuration;
   i++)
  {
    int pirValue =
   analogRead(pirPin);
    if (pirValue > 10)
    {
      digitalWrite(buzzPlusPin, HIGH);
      alarmTimeout = alarmDuration;
    }
    if (alarmTimeout <= 0)
    {
      digitalWrite(buzzPlusPin, LOW);
    }
    alarmTimeout -;
    delay(1000);
  }
}

void moveToNewPlace()
{
  turnInRandomDirection();
  forwardOrProximity(1500);
}

void turnInRandomDirection()
{
  int duration = random(100, 3000);
  left();
  delay(duration);
  halt();
}

void forwardOrProximity(int duration)
{
  int x = 0;
  forward();
  static float lastProx = 0;
  float prox = 0;
  while (x < duration)
  {
    int rawProx = analogRead(proxPin);
    prox = alpha * rawProx + (1 -
     alpha) * lastProx;
    Serial.print(rawProx);
    Serial.print(" ");
    Serial.print(lastProx);
    Serial.print(" ");
    Serial.println(prox);
```

LISTING 15-1 *(continued)*

```
    lastProx = prox;
    if (prox > proxThreshold)
    {
      halt();
      buzz(50); buzz(50);
      delay(100);
      back();
      delay(700);
      halt();
      return;
    }
    x += 10;
    delay(10);
  }
}

void forward()
{
 setLeft(CLOCKWISE);
 setRight(CLOCKWISE);
}

void back()
{
 setLeft(COUNTER_CLOCKWISE);
 setRight(COUNTER_CLOCKWISE);
}

void left()
{
 setLeft(CLOCKWISE);
 setRight(COUNTER_CLOCKWISE);
}

void right()
{
 setLeft(COUNTER_CLOCKWISE);
 setRight(CLOCKWISE);
}

void halt()
{
 setLeft(HALT);
 setRight(HALT);
}

void setLeft(int rotation)
{
 if (rotation == HALT)
 {
   digitalWrite(leftAPin, LOW);
```

(continued on next page)

LISTING 15-1 *(continued)*

```
    digitalWrite(leftBPin, LOW);
  }
  else if (rotation == CLOCKWISE)
  {
    digitalWrite(leftAPin, HIGH);
    digitalWrite(leftBPin, LOW);
  }
  else if (rotation ==
    COUNTER_CLOCKWISE)
  {
    digitalWrite(leftAPin, LOW);
    digitalWrite(leftBPin, HIGH);
  }
}

void setRight(int rotation)
{
  if (rotation == HALT)
  {
    digitalWrite(rightAPin, LOW);
    digitalWrite(rightBPin, LOW);
  }
  else if (rotation == CLOCKWISE)
  {
    digitalWrite(rightAPin, HIGH);
    digitalWrite(rightBPin, LOW);
  }
  else if (rotation ==
    COUNTER_CLOCKWISE)
  {
    digitalWrite(rightAPin, LOW);
    digitalWrite(rightBPin, HIGH);
  }
}

void buzz(int duration)
{
  digitalWrite(buzzPlusPin, HIGH);
  delay(duration);
  digitalWrite(buzzPlusPin, LOW);
  delay(duration);
}
```

two sensors connected to analog inputs of the Arduino.

We then have a section of variables that we may be interested in tweaking to modify the behavior of our robot:

- monitorDuration determines how long in seconds the robot should stay still while monitoring before it moves to a new location.

- alarmDuration sets the period in seconds the alarm should sound once movement is detected. You will probably want to keep this short during testing!

The "setup" function initializes all these pins to the right mode. You will also see that we set posPin and negPin to 5V and GND, respectively. We are actually using those pins to supply power to the proximity sensor. This lets us use adjacent pins in the wiring, thus simplifying matters.

One of the things we have to be very careful about is to prevent both the A and B control signals to a motor controller from becoming high at the same time. If this happens, there will be smoke and dead transistors! To prevent this, we limit the places in the sketch where we change those pins to the functions setLeft and setRight. These functions set the direction of turning for a motor to be one of HALT, CLOCKWISE, or COUNTER_CLOCKWISE. If you always use these functions, you can't go wrong.

In a layer above these functions are another series of functions—halt, forward, backward, left, and right—that control the direction of the robot as a whole rather than the individual motors.

At the top of the heap, we have the "loop" function, which is invoked continuously. This first calls the "monitor" function and then the moveToNewPlace function.

The "monitor" function repeatedly reads the analog input connected to the PIR sensor, and if it is over the trigger threshold, it sets a timer and turns on the buzzer. When the timer has expired, the buzzer is turned back off again.

The moveToNewPlace function invokes the two methods called turnInRandomDirection and forwardOrProximity in turn. As their names suggest, turnInRandomDirection rotates to the left for a random period of between 100 and 3000 milliseconds and forwardOrProximity moves forward for the time specified in its argument, or until the proximity sensor detects an obstacle. If an obstacle is detected, the robot buzzes twice and then reverses.

IR Distance Sensor

The IR distance sensor uses triangulation to detect the distance of an object (Figure 15-19).

A bright IR LED with a focusing lens projects a dot. This will be reflected by any object, and the reflected light then falls on a linear CCD array of photocells. The most brightly illuminated cells will vary according to the distance of the reflected object. The control electronics then provide an analog voltage that indicates the distance.

One side effect of the way this works is that the sensor has both a maximum and minimum distance it can sense. So, if the robot becomes too close to an object, it will not be able to tell that an object is close.

Another effect of this design is that the sensor is not linear, but rather follows a parabolic curve. Fortunately for us, we just want to know when we are close to an obstacle, not how far away it is.

Driving Motors

Most robots need motors of some description and many will need to drive them from a microcontroller like the Arduino. Just turning a motor on and off is quite straightforward. MOSFET transistors (see the "Theory" section of Chapter 5) are often used for this because of their excellent switching characteristics—that is, very low "on" resistance and very high "off" resistance. They also take very little current to turn them on and off.

Figure 15-20 shows the basic schematic for controlling a motor using a single N-channel MOSFET.

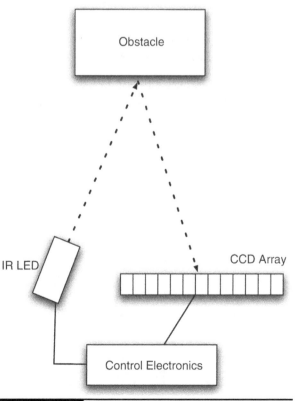

Figure 15-19 IR distance sensor

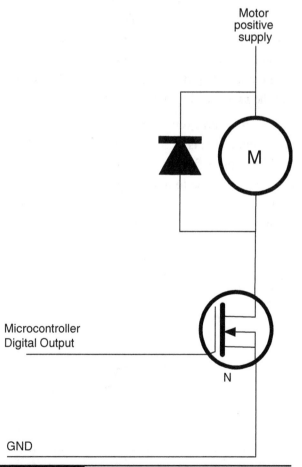

Figure 15-20 Controlling a motor with a MOSFET

The diode is there to protect the MOSFET against the large back emf currents that can flow when driving inductive loads like motors.

Life gets more complicated when we want to be able to drive the motor in both directions. To reverse the direction of a DC motor, just send the current through it in the opposite direction. That is, you reverse the polarity of the voltage across its terminals, so the side that was connected to ground becomes connected to the positive supply, and vice versa.

The normal approach to switching the current direction like this is to use something called an H-bridge. Figure 15-21 shows the operation of an H-bridge using four switches. In reality, you would either use four MOSFETs or an IC H-bridge.

In Figure 15-21, S1 and S4 are closed, and S2 and S3 are open. This allows current to flow through the motor, with terminal A positive and terminal B negative. If we were to reverse the switches so that S2 and S3 are closed and S1 and S4 are open, then B will be positive and A negative and the motor will turn in the opposite direction.

You may, however, have spotted a danger with this circuit. That is, if by some chance S1 and S2 are both closed, then the positive supply will be directly connected to the negative supply and a short circuit will result. The same is true if S3 and S4 are both closed at the same time.

In this project, we sort of use half an H-bridge, because we have a split power supply, so one side of the motor is always connected to the middle voltage from the battery, and to get it to move in either direction, all we have to do is make the other connection connect to either the positive supply or ground. This is shown in Figure 15-22 in a simplified form using switches.

As with a full H-bridge, you can see that if both S1 and S2 are closed at the same time, there will be a short circuit across the battery. If S1 and S2 are transistors, then the resulting high current has nothing much to limit it and the transistors will probably be fried.

Some H-bridge implementations, especially those that use an IC, will include built-in protection so a software error cannot result in a short circuit.

Summary

Arduino boards make great controllers for a robot. They are relatively easy to program and have a good selection of input/output pins. This is a very flexible design, which could lend itself to many possibilities.

Some things you could do with your robot base are:

- Attach a WiFi webcam to it

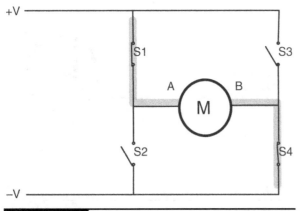

Figure 15-21 An H-bridge

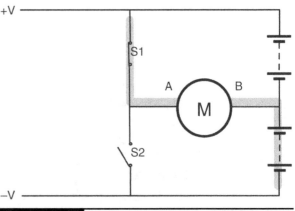

Figure 15-22 Half an H-bridge

- Remote control it using Bluetooth
- Attach a mini vacuum cleaner to it and make your own autonomous hovering robot
- Arm it with the coil gun from Chapter 2

This book should provide the Evil Genius with all the necessary technology to achieve world domination. Having imparted his knowledge to fledgling Evil Geniuses all over the world, it just remains for the Evil Genius to insist that his followers direct themselves to the website www.dangerouslymad.com. Here, you will find further information about the book, and useful resources such as the message designer for the persistence-of-vision display.

Electronics Construction Primer

THIS APPENDIX PROVIDES a few useful hints and tips for electronic construction used in the majority of the projects in this book.

Circuits

The author likes to start a project with a vague notion of what he wants to achieve and then start designing from the perspective of the electronics.

The way to express an electronic circuit is to use a schematic diagram. The author has included schematic diagrams for all the projects in this book, so even if you are not very familiar with electronics, you should now have seen enough schematics to understand roughly how they relate to the stripboard diagrams also included.

Schematic Diagrams

In a schematic diagram, connections between components are shown as lines. These connections will use the copper strips of a piece of stripboard and the wires connecting one strip to another.

Schematic diagrams (Figure A-1) have a few conventions that are worth pointing out. For instance, it is common to place GND lines near the bottom of a diagram and higher voltages near the top of a diagram. This allows someone reading the schematic to visualize the flow of charge through the system from higher voltages to lower voltages.

When there is not enough room to draw all the connections to GND, another convention in schematic diagrams is to use the little bar symbol to indicate a connection to GND.

Schematic diagrams can be drawn with many different tools. Some of them are integrated electronics CAD (computer-aided design) products that will go on to lay out the tracks on a printed circuit board for you. By and large, these create fairly ugly-looking diagrams, and because of this the author prefers to use pencil and paper or general-purpose drawing software. All the diagrams for this book were created using Omni Group's excellent but strangely named OmniGraffle software, which is only available for Macs. OmniGraffle templates for drawing breadboard layouts and schematic diagrams are available for download from www.dangerouslymad.com.

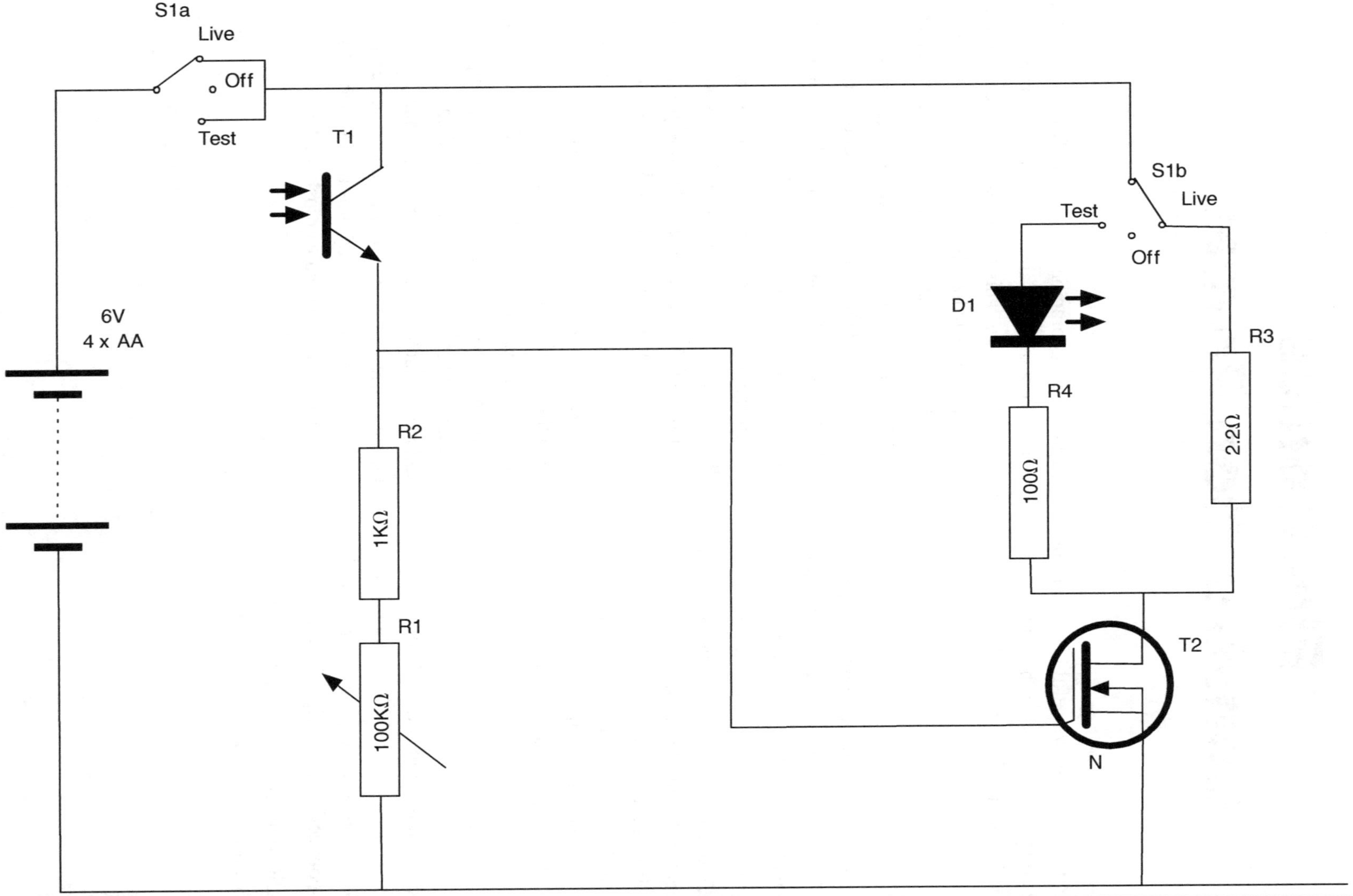

Figure A-1 A schematic diagram example

Component Symbols

Figure A-2 shows the circuit symbols for some common electronics.

Various standards exist for circuit diagrams, but the basic symbols are all recognizable between standards. The set used in this book does not closely follow any particular standard. I have just chosen what I consider to be the most easy-to-read approach to the diagrams.

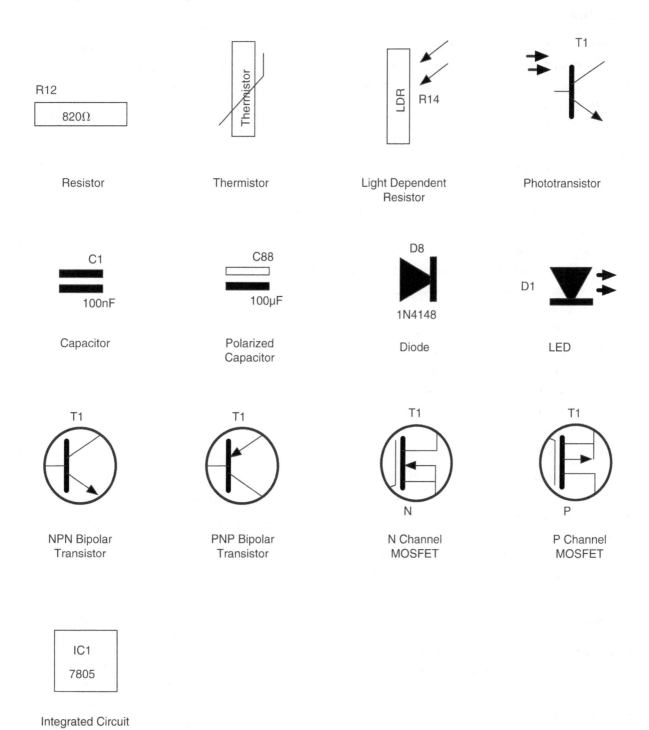

Resistor	Thermistor	Light Dependent Resistor	Phototransistor
Capacitor	Polarized Capacitor	Diode	LED
NPN Bipolar Transistor	PNP Bipolar Transistor	N Channel MOSFET	P Channel MOSFET

Integrated Circuit

Figure A-2   Circuit symbols

Components

In this section, we look at the practical aspects of components: what they do, and how to identify, choose, and use them.

Datasheets

All component manufacturers produce datasheets for their products. These act as a specification for how the component will behave. They are not of much interest for resistors and capacitors, being much more useful for semiconductors and transistors, and especially integrated circuits. They will often include application notes that contain example schematics for using the components.

These are all available on the Internet. However, if you search for "BC158 datasheet" in your favorite search engine, you will find many of the top hits are for organizations cashing in on the fact that people search for datasheets a lot. These organizations surround the datasheets with pointless advertising and pretend they add some value to looking up datasheets by subscribing to their service. In reality, such web sites usually just lead to a frustration of clicking and should be ignored in favor of any manufacturer's web sites. So, scan down the search results until you see a URL like www.fairchild.com.

Alternatively, many of the component retail suppliers such as Farnell provide free-of-charge and nonsense-free datasheets for pretty much every component they sell, which is to be much applauded. It also means that you can price and buy the components while finding out more about them.

Resistors

Resistors are the commonest and cheapest electronic components around. Their typical uses are:

- To prevent excessive current flowing (see any projects that use an LED)

- In a pair or as a variable resistor, to divide a voltage

In the "Theory" section of Chapter 12, we looked at Ohm's law and used it to decide on a value of series resistor for an LED.

Resistors have colored bands around them to indicate their values. However, if you are unsure of a resistor, you can always find its resistance using a multimeter. Once you get the hang of it, it's easy to read the values using the colored bands.

Each band color has a value associated with it, as shown in Table A-1.

TABLE A-1	Resistor Color Codes
Color	**Value**
Black	0
Brown	1
Red	2
Orange	3
Yellow	4
Green	5
Blue	6
Violet	7
Gray	8
White	9

There will generally be three of these bands together, starting at one end of the resistor, which is then followed by a gap, and finishes with a single band at the other end of the resistor. The single band indicates the accuracy of the resistor value. Since none of the projects in this book require very accurate resistors, there is no need to select your resistors on the basis of accuracy.

Figure A-3 shows the arrangement of the colored bands. The resistor value uses just the three bands. The first band is the first digit, the second the second digit, and the third "multiplier"

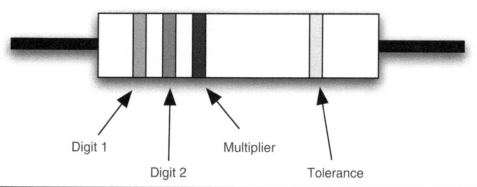

Digit 1 Multiplier

Digit 2 Tolerance

Figure A-3 A color-coded resistor

band is how many zeros to put after the first two digits.

So a 270Ω resistor will have first digit 2 (red), second digit 7 (violet), and a multiplier of 1 (brown). Similarly, a 10kΩ resistor will have bands of brown, black, and orange (1, 0, 000).

Most of our projects use resistors in a very low-power manner. A quick calculation can be used to work out the current flowing through the resistor, and multiplying that by the voltage across it will tell you the power used by the resistor. The resistor burns off this surplus power as heat, so resistors will get warm if a significant amount of current flows through them.

You only need to worry about this for low-value resistors of less than 100Ω or so, because higher values will have such a small current flowing through them.

As an example, a 100Ω resistor connected directly between 5V and GND will have a current through it of $I = V/R$ or 5/100, or 0.05 amps.

The power it will use will be $I \times V$ or $0.05 \times 5 = 0.25$ W.

A standard power rating for resistors is 0.5 W or 0.6 W, and unless otherwise stated in the projects, 0.5-W metal film resistors will be fine.

Transistors

Browse through any component catalog and you will find literally thousands of different transistor types. In this book, the list has been simplified to the entries in Table A-2.

If it proves difficult to find a particular component, you can usually find alternatives by comparing datasheets.

The basic switch circuit for a transistor is shown in Figure A-4.

The current flowing from base to emitter (b to e) controls the larger current flowing from the collector to the emitter. If no current flows into the base, then no current will flow through the load. In

TABLE A-2	Transistors Used in This Book	
Transistor	**Type**	**Purpose**
BC548	Bipolar NPN	Switching small loads greater than 40mA
2N7000	N-channel FET	Low-power switching with on resistance (see Project 7)
FQP33N10	N-channel power MOSFET	High-power switching
FQP27P06	P-channel power MOSFET	High-power switching

+V

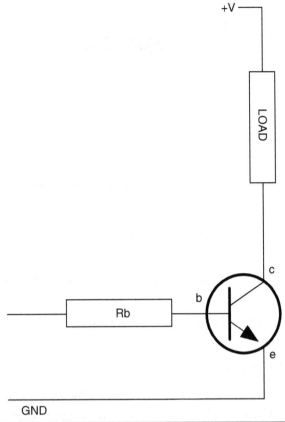

LOAD

c

b

Rb

e

GND

Figure A-4   Basic transistor switch circuit

most transistors, if the load has zero resistance, the current flowing into the collector will be 50 to 200 times the base current. However, we will be switching our transistor fully on or fully off, so the load resistance will always limit the collector current to the current required by the load.

Too much base current will damage the transistor and also rather defeat the objective of controlling a bigger current with a smaller one. So, the base will have a resistor connected to it.

When switching from an Arduino board, the maximum current of an output is 40mA, so we could choose a resistor that allows about 30mA to flow when the output pin is high at 5V. Using Ohm's law:

$$R = V/I$$

$$R = (5 - 0.6)/30 = 147\Omega$$

The "– 0.6" is because one characteristic of bipolar transistors is that there is always a voltage

of about 0.6V between the base and emitter when a transistor is turned on.

Using a 150Ω base resistor, we could therefore control a collector current of 40 to 200 times 30mA, or 1.2A to 6A, which is more than enough for most purposes.

In practice, we would probably use a resistor of 1kΩ, or perhaps 270Ω.

Transistors have a number of maximum parameter values that should not be exceeded, otherwise the transistor may be damaged. You can find these by looking at the datasheet for the transistor.

For example, the datasheet for a BC548 will contain many values. The ones of most interest to us are summarized in Table A-3.

TABLE A-3	Transistor Datasheet	
Property	**Value**	**What It Means**
I_c	100mA	The maximum current that can flow through the collector without the transistor being damaged.
h_{FE}	110–800	DC current gain. This is the ratio of collector current to base current, and as you can see, it can be anything between 110 and 800 for this transistor.

Diodes

In addition to LEDs, there are also "normal" diodes. These act a bit like one-way valves, only allowing current to flow through them in one direction.

We have used them in a few of the projects in this book to prevent unwanted currents from damaging components, or when driving inductive loads like the coil in the levitator or the motors in the surveillance robot.

This "one-way" property of diodes is also what allows them to rectify a signal (see the "Theory" section of Chapter 9).

Buying Components

Thirty years ago, the electronic enthusiast living in even a small town would be likely to have the choice of several radio/TV repair shops where they could buy components and receive friendly advice. These days, a few retail outlets still sell components, like RadioShack in the U.S. and Maplin's in the UK, but the Internet has stepped in to fill the gap, making it easier and cheaper than ever to buy components.

With international component suppliers such as RS and Farnell, you can fill a virtual shopping basket online and have the components arrive in a day or two. Shop around, because prices vary considerably between suppliers for the same components.

You will find eBay to be a great source of components. Also, if you don't mind waiting a few weeks for your components to arrive, great bargains can be had from China. You often have to buy large quantities, but you may find it cheaper to get 50 of a component from China than 5 locally. That way, you have some spares for your component box.

Tools

When making your own projects, the following tools will be needed as a bare minimum.

- Multi-core wire in a few different colors; something around 0.6mm (23 SWG) wire diameter
- Pliers and wire snips
- A multimeter

- A soldering iron
- Solder
- Assorted screwdrivers

Component Boxes

When you first start designing your own projects, it will take you some time to gradually build up your stock of components. Each time you are finished with a project, a few more components will find their way back to your stock.

It is useful to have a basic stock of components so you don't have to keep ordering things when you just need a different-value resistor. You have probably noticed that most of the projects in this book tend to use values of resistors like 100Ω, $1k\Omega$, $10k\Omega$, and so on. You actually don't need that many different components to cover most of the bases for a new project.

Boxes with compartments, which can be labeled, save a lot of time in selecting components, especially when it comes to resistors that don't have their values written on them.

Snips and Pliers

Snips are for cutting, and pliers are for holding things still (often while you cut them).

Figure A-5 shows how you strip the insulation off wire. Assuming you are right-handed, hold

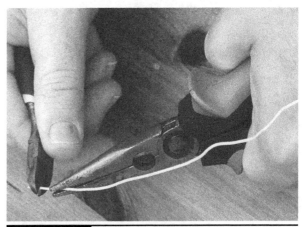

Figure A-5   Snips and pliers

your pliers in your left hand and the snips in the right. Grip the wire with the pliers close to where you want to start stripping the wire from, and then gently pinch around the wire with the snips, pulling sideways to strip the insulation away. Sometimes you will pinch too hard and cut or weaken the wire, and other times you will not pinch hard enough and the insulation will remain in tact. It's all just a matter of practice.

You can also buy an "automatic" wire stripper that grips and removes insulation in one action. In practice, these often only work well for one particular wire type, and sometimes just plain don't work.

Soldering

You do not have to spend a lot of money to get a decent soldering iron. Temperature-controlled solder stations are better (Figure A-6), but a fixed-temperature mains electric iron is fine. Buy one with a fine tip and make sure it's the kind intended for use with electronics, not plumbing.

Use narrow lead-free solder. Anyone can solder things together and make them work; however, some people just have a talent for neat soldering. Don't worry if your results don't look as neat as a machine-made printed circuit. They never will.

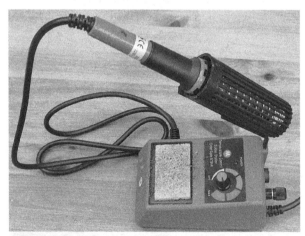

Figure A-6   A soldering iron and solder

Soldering is one of those jobs that you really need three hands for. One hand to hold the soldering iron, one to hold the solder, and one to hold the thing you are soldering. Sometimes the thing you are soldering is big and heavy enough to stay put while you solder it; other times, you will need to hold it down. Heavy pliers are good for this, as are mini vices and "helping hand"–type holders that use little clips to grip things.

The basic steps for soldering are:

- Wet the sponge in the soldering iron stand.
- Allow the iron to come up to temperature.
- Tin the tip of the iron by pressing solder against it, until it melts and covers the tip.
- Wipe the tip on the wet sponge—this produces a very satisfying sizzling sound, but also cleans off the excess solder. You should now have a nice bright silver tip.
- Touch the iron to the place where you are going to solder (in order to heat it), then after a short pause (a second or two) touch the solder to the point where the tip of the iron meets the thing you are soldering. The solder should flow like a liquid, neatly making a joint.
- Remove the solder and soldering iron, putting the iron back in its stand, and being very careful that nothing moves in the few seconds that the solder takes to solidify. If something does move, touch the iron to it again to re-flow the solder; otherwise, you can get a bad connection called a "dry joint."

Above all, try not to heat sensitive (or expensive) components any longer than necessary, especially if they have short leads.

Practice soldering old bits of wire together or solder wires to an old section of circuit board before working on the real thing.

Multimeters

A big problem with electrons is that you cannot see them. A multimeter lets you view what they are up to. It allows you to measure voltage, current, resistance, and often other features, too, like capacitance, frequency, and more. A cheap $10 multimeter is perfectly adequate for most purposes.

Multimeters (Figure A-7) can be either analog or digital. You can tell more from an analog meter than you can from a digital, because you can see how fast a needle swings over, and how it jitters—something that isn't possible with a digital meter, where the numbers just change. However, for a steady voltage, it's much easier to read a digital meter since an analog meter will have a number of scales, and you have to work out which scale you should be looking at before taking the reading.

You can also get autoranging meters which, once you have selected whether you are measuring current or voltage, will automatically change ranges for you as the voltage or current increases. This is useful, but some would argue that thinking about the range of voltage before you measure it is actually a good thing.

To measure voltage using a multimeter:

■ Set the multimeter range to voltage (start at a range that you know will be higher than the voltage you are about to measure).

■ Connect the black lead to GND. A crocodile clip on the negative lead makes this easier.

■ Touch the red lead to the point whose voltage you want to measure. For instance, to see if an Arduino digital output is on or off, you can touch the red lead to the pin and read the voltage, which should be either 5V or 0V.

Measuring current is different than measuring voltage because you want to measure the current flowing through something, not the voltage at a particular point. So you put the multimeter in the path of the current you are measuring. This means that when the multimeter is set to a current setting, there will be a very low resistance between the two leads, so be careful not to short anything out with the leads.

Figure A-8 shows how you could measure the current flowing through an LED.

To measure current:

■ Set the multimeter range to a current range higher than the expected current. Note that multimeters often have a separate high-current connector for currents as high as 10A.

■ Connect the positive lead of the meter to the more positive side from which the current will flow.

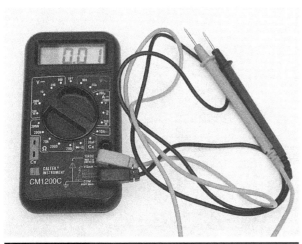

Figure A-7 A multimeter

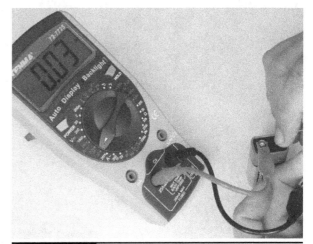

Figure A-8 Measuring current

- Connect the negative lead of the meter to the more negative side. Note that if you get this the wrong way around, a digital meter will just indicate a negative current, but if you've connected an analog meter the wrong way around, it may damage it.

- In the case of an LED, the LED should still light as brightly as before you put the meter into the circuit, letting you read the current consumption.

Another feature of a multimeter that is sometimes useful is the continuity test feature. This will usually beep when the two test leads are connected together. You can use this to test fuses and so on, but also to test for accidental short circuits on a circuit board, or broken connections in a wire.

Resistance measurement is occasionally useful, particularly if you want to determine the resistance of an unmarked resistor.

Some meters also have diode and transistor test connections, which can be useful in finding and discarding transistors that have burned out.

Oscilloscopes

Oscilloscopes (Figure A-9) are an indispensable tool for any kind of electronics design or test where you are looking at a signal that changes over time. They are a relatively expensive bit of equipment and there are various types. One of the most cost-effective types does not have any display at all, but connects to your computer over USB. If you don't want to risk blobs of solder on your laptop, or wait for it to boot up, then a dedicated oscilloscope is probably the answer for you.

Entire books have been written about using an oscilloscope effectively, and since every oscilloscope is different, we will just cover the basics here.

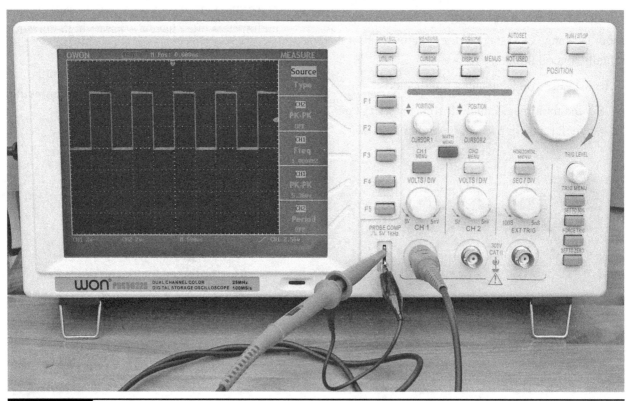

Figure A-9   Oscilloscope

As you can see from Figure A-9, the waveform is displayed over the top of a grid. The vertical grid is in units of some fraction of volts, which on this screen is 2V per division. So the voltage of the square wave in total is 2.5 × 2 or 5V.

The horizontal axis is the time axis, which is calibrated in seconds—in this case, 500 microseconds (mS) per division. So the length of one complete cycle of the wave is 1000 mS, or 1 millisecond, indicating a frequency of 1 kHz.

Summary

Many resources are available to help you learn electronics. On the Internet, electronics forums abound, where you can post your questions and receive high-quality answers.

Books and hobby electronics magazines offer you useful projects with some words of explanation about their designs. As time goes on, you will find yourself understanding more and more.

Also, seek out clubs in your area. Radio amateur "Ham Fests" (nothing to do with pork) are a great way to meet other enthusiasts and pick up interesting bargain components.

Index

References to figures are in italics.

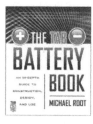

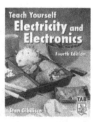